ANTIOXIDANT NUTRIENTS AND IMMUNE FUNCTIONS

ADVANCES IN EXPERIMENTAL MEDICINE AND BIOLOGY

Recent Volumes in this Series

A Continuation Order Plan is available for this series. A continuation order will bring delivery of each new volume immediately upon publication. Volumes are billed only upon actual shipment. For further information please contact the publisher.

ANTIOXIDANT NUTRIENTS AND IMMUNE FUNCTIONS

Edited by

Adrianne Bendich

Hoffmann LaRoche, Inc.
Nutley, New Jersey

Marshall Phillips

Agricultural Research Service, USDA, NADC
Ames, Iowa

and

Robert P. Tengerdy

Colorado State University
Fort Collins, Colorado

SPRINGER SCIENCE+BUSINESS MEDIA, LLC

Library of Congress Cataloging in Publication Data

Agricultural and Food Chemistry Division of the American Chemical Society Symposium on Antioxidant Nutrients and the Immune Response (1988: Los Angeles, Calif.)
    Antioxidant nutrients and immune functions / edited by Adrianne Bendich, Marshall Phillips, Robert P. Tengerdy.
        p.   cm. — (Advances in experimental medicine and biology; v. 262)
    Includes bibliographical references.
    ISBN 978-1-4612-7863-4   ISBN 978-1-4613-0553-8 (eBook)
    DOI 10.1007/ 978-1-4613-0553-8
    1. Immunity—Nutritional aspects—Congresses. 2. Antioxidants—Physiological effect—Congresses. 3. Active oxygen—Physiological effect—Congresses. I. Bendich, Adrianne. II. Phillips, Marshall. III. Tengerdy, Robert P. IV. American Chemical Society. Division of Agricultural and Food Chemistry. V. Title. VI. Series.
    QR185.2.A36 1988   89-26570
    599'.029—dc20   CIP

Proceedings of the Agricultural and Food Chemistry Division of
the American Chemical Society Symposium on Antioxidant Nutrients
and the Immune Response, held September 29, 1988,
in Los Angeles, California

© 1990 Springer Science+Business Media New York
Originally published by Plenum Press, New York in 1990
Softcover reprint of the hardcover 1st edition 1990

Adrianne Bendich dedicates this book to:
David, Jorden, and Debra,
Joseph, Lillian, and Elaine

Marshall Phillips dedicates this book to:
Karen, Tacy, Dori, and Larry

Robert Tengerdy dedicates this book to:
Katherine, Thomas, and Peter

PREFACE

The determination of optimal nutritional status has traditionally been based upon generalized parameters such as weight gain and body fat levels. Vitamin and mineral requirements were often related to the intakes needed to prevent overt signs of deficiency diseases such as beriberi or scurvy. However, in the past decade or so, there have been intensive investigations to determine the subtle changes in physiological functions associated with marginal micronutrient intakes. There is a growing consensus that immune system activities are very sensitive indicators of micronutrient status. During this decade, there has also been a rapid expansion of research in the role of free radicals and antioxidants in the major chronic diseases which afflict mankind(i.e. cancer, cardiovascular disease,and autoimmune disease).

The main function of antioxidant nutrients in an appropriate diet is the prevention of oxidative damage to cells and their physiological functions. Antioxidant nutrients counteract free radicals and damaging oxidative actions on cell membranes. Since the cells of the immune system are rapidly differentiating and proliferating, such dividing and transforming cells are particularly susceptible to damage by oxidation.

The interactions of antioxidant nutrition and immune system activities and disease resistance are therefore logical areas for research. Thus, the objective of this symposium was to bring together the leading investigators who have examined the immunological effects of dietary essential nutrients which share the capacity to act as antioxidants.

Over the past 15 years, there has been a growing interest in the reproducible adverse effects of the deficiencies of vitamins C and E, zinc, selenium or copper deficiency of cellular and humoral responses as well as phagocytic function. Oxidative damage to leukocytes and immunosuppressive effects of prostaglandin E 2, lipid peroxides and conjugated immunoglobulins are possible rationales for immunosuppression associated with antioxidant micronutrient deficiencies. Similarly, many studies have also found immunoenhancement associated with dietary supplementation of the above micronutrients as well as with beta carotene.

Sound nutrition consists of a necessary balance between the various elements of the diet: protein, calories, lipids, vitamins and minerals. The nutritional state of individuals may be a clue that determines immunocompetence and disease resistance.The interest in proper nutrition has moved beyond the scientific laboratory and has reached into the public awareness. The perception of the health consequences of sound nutrition is as important as the scientific validity. Therefore, it is anticipated that the research from the emerging field of nutritional immunology will be seriously considered in the development of nutrition policy. Because of the importance of antioxidant nutritional status and its effects on immunocompetence, further research is recommended with emphasis on clinical and epidemiological areas.

Adrianne Bendich
Marshall Phillips
Robert Tengerdy

ACKNOWLEDGEMENTS

The authors acknowledge the Division of Agricultural and Food Chemistry of the American Chemical Society for sponsoring this symposium at the National meeting in Los Angeles.

Financial contributions and support from Hoffman-LaRoche, Inc., Mead Johnson Nutritional Division, Nabisco Brands, Inc., Quaker Oats Company, and Takeda U.S.A., Inc., are sincerely acknowledged.

We appreciate the technical assistance of Ms. Janice Olson whose dedication to the production of this book is greatly appreciated.

CONTENTS

ANTIOXIDANT NUTRIENTS AND IMMUNE FUNCTIONS - INTRODUCTION

Adrianne Bendich

Clinical Research Coordinator
Hoffmann-LaRoche Inc.
340 Kingsland St.
Nutley, N.J.

INTRODUCTION

The purpose of this chapter is to examine the role of antioxidants, and thus the role of free radicals in the activities of the immune system. The chapter is divided into three main sections, free radicals, antioxidants and an introduction to the activities of the immune system, especially those involving oxidative reactions. The information presented by each symposium speaker is highlighted in this introduction so that the reader can refer to the appropriate chapter for more detailed information.

FREE RADICALS

Chemical compounds contain two or more elements that are bound together by a chemical bond. In most instances, the bonding involves negatively-charged electrons. The arrangement of the electrons determine the stability of the compound. A stable compound has electrons that are paired. If an electron is unpaired, the molecule becomes more reactive and unstable than the parent compound (Dormandy, 1983). A compound or element with one or more unpaired electrons is called a free radical.

In order to stabilize itself, the free radical abstracts an electron from a stable compound which, in turn, is transformed into a new free radical. This chain reaction will continue until the free radical containing the lone electron pairs up with another molecule with an unpaired electron, or is deactivated by a chain reaction-breaking antioxidant. There are also certain free radical scavengers and enzymes which can facilitate the decomposition of reactive molecules which are precursors of free radicals. These compounds can therefore lower the free radical burden before a chain reaction begins (Grisham and McCord, 1986; Sies, 1985).

Oxygen containing reactive molecules (some of which are free radicals) are the most critical to biological systems. Since oxygen is required for cell viability, it is essential that mechanisms are available to control the reactive oxygen intermediates generated during cellular respiration. As seen in Fig 1, there are three major reactions of oxygen which result in the formation of oxygen species more reactive than molecular oxygen.

When oxygen is exposed to a source of high energy such as ultraviolet light, the energy is transferred to form singlet oxygen (reaction 1). Singlet oxygen still contains a pair of electrons and cannot be classified as a free radical. However, these two electrons exist in an unstable conformation in the molecule and thus the potential exists for this molecule to participate in reactions that generate free radicals. One example of a free radical produced through singlet oxygen reactions is the superoxide radical. Singlet oxygen can be deactivated, thereby

1

preventing the formation of the superoxide radical. This process is called singlet oxygen quenching ((Halliwell and Gutteridge, 1985). The quencher dissipates energy contained in the singlet oxygen molecule. As a result, it lacks the energy to engage in superoxide-producing reactions. The most effective, naturally-occurring singlet oxygen quencher is beta-carotene (Krinsky and Deneke, 1982).

1) O_2 + energy $\longrightarrow$ $O_2^{\bullet}$ (singlet oxygen)

2) O_2 + 4 e^- + 4 H^+ $\longrightarrow$ 2 H_2O (mitochondrial electron transport system)

3a) O_2 + e^- $\longrightarrow$ $O_2^{\bullet -}$ (superoxide radical anion)

3b) 2 $O_2^{\bullet -}$ + 2 H^+ $\longrightarrow$ O_2 + H_2O_2 (hydrogen peroxide)

3c) O_2 + 2 e^- + 2 H^+ $\longrightarrow$ H_2O_2

3d) $O_2^{\bullet -}$ + H_2O_2 + 4 $\overset{\bullet}{H}{}^+$ $\longrightarrow$ O_2 + H_2O + $HO^{\bullet}$ (hydroxyl radical)

Fig. 1. Reactions with oxygen resulting in more reactive species than oxygen.

Over 98% of oxygen utilized by cells is efficiently reduced by the electron transport system (reaction 2). However the remaining oxygen is available for one or two electron reduction, resulting in the formation of the superoxide radical (reaction 3a) and hydrogen peroxide (reaction 3b or 3c). These two products can react to form water, however, in the process, the highly reactive hydroxyl radical is generated (3d). Electron transport systems are thus prime, continuous sources of intracellular, reactive oxygenated free radicals. In addition, electron transfer from the superoxide radical to transition metals such as iron also generates hydroxyl radicals (Grisham & McCord, 1986).

The cellular sites of free radical generation include mitochondria, lysosomes, peroxisomes, nuclear, endoplasmic reticular and plasma membranes. Free radicals and their products are also present within the cytosol. Molecular species include, but by no means are limited to hydroxyl, peroxy, hypochlorite, superoxide and alkoxy radicals and reactive molecules such as hydrogen peroxide and singlet oxygen, which are not free radicals but are certainly reactive and capable of causing damage (Machlin and Bendich, 1987).

Prime targets for free radical reactions are the unsaturated bonds in the lipids found in all cellular membranes. Consequently, there is loss of membrane fluidity, receptor alignment and potentially cellular lysis. Free radical damage to the sulfur containing enzymes and other proteins results in inactivation, cross-linking and denaturation. Nucleic acids can be attacked and subsequent damage to the DNA can cause mutations which may be carcinogenic. Oxidative damage to carbohydrates can alter any of the cellular receptor functions including those associated with interleukin activities and prostaglandin formation. Examples of tissue and organ damage associated with free radical reactions include the destruction of endothelial cells which line the blood vessels and consequent macrophage invasion (Halliwell and Gutteridge, 1985; Flohe et al., 1985; Gey et al., 1987), the inflammatory response seen in arthritic joints (Hirschelmann and Bekemeier, 1981; Slater, 1972), the hastening of the aging process (Cutler, 1986), and the destruction of lung tissue (Freeman and Crapo, 1982).

Endogenous sources of oxygen-containing free radicals include those generated intracellularly as well as those that are formed within the cell and released into the surrounding area. Intracellular free radicals are generated from the autooxidation and consequent inactivation of small molecules such as thiols and catecholamines, the activity of certain oxidases,

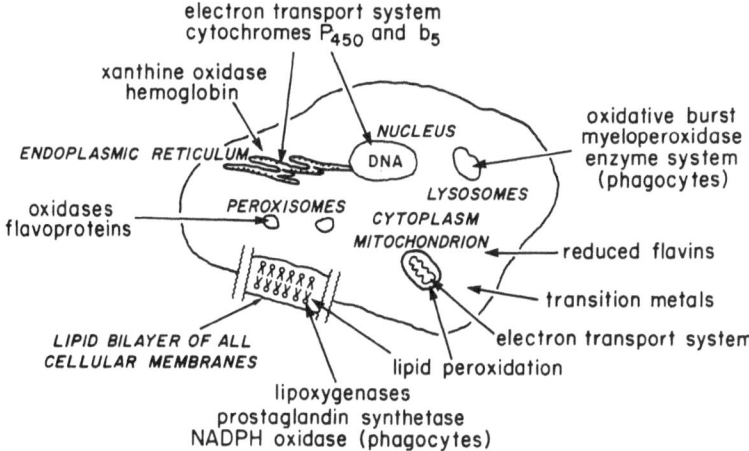

Fig. 2. Cellular sources of free radicals.

cyclooxygenases, lipoxygenases, dehydrogenases, peroxidases and several other metabolizing pathways (Fig. 2.; Freeman and Crapo, 1982).

In addition to the continuous production of reactive oxygen intermediates during normal cellular metabolism, the process of killing bacteria and other pathogens which invade the body utilizes free radicals. The major cell type associated with this first line of defense against infection is the neutrophil. Neutrophils have the capacity to take up molecular oxygen and generate reactive oxygen-containing molecules when stimulated. This is often called the oxidative or respiratory burst. Free radicals and singlet oxygen, along with other reactive molecules, can kill bacterial pathogens. Neutrophils can also generate highly toxic halogenated molecules when the myeloperoxidase halide enzyme system is activated during the oxidative burst. The halogenated species can also lyse pathogens. The killing process is usually confined to intracellular vacuoles which enclose the phagocytized pathogen (Anderson, 1982).

Exogenous sources of free radicals include radiation, tobacco smoke, certain pollutants and organic solvents, anesthetics, hyperoxic environments and pesticides. Some of these compounds as well as certain medications are metabolized to free radical intermediate products which have been shown to cause oxidative damage to the target tissues such as the liver and the lung (Slater, 1987).

ANTIOXIDANTS

Although there are numerous sources of free radicals, there are also several naturally occurring compounds that can inactivate these reactive molecules. Antioxidants have the capacity to lower the free radical burden. Free radical reactions can be broken down into three stages; initiation, propagation and termination. Antioxidants can affect the generation of free radicals during any of these stages.

Mineral-Containing Antioxidant Enzymes

Two important antioxidants are metalloenzymes which can interfere with the production of free radicals during the initiation phase by inactivating precursor molecules. There are two types of superoxide dismutases; a Mn-containing enzyme in mitochondria, and a Cu/Zn-containing enzyme in the cytoplasm, both of which catalyze the reaction seen in Fig 3. Catalase, an Fe-containing enzyme found in peroxisomes, catalyzes the decomposition of the hydrogen peroxide produced as a result of superoxide dismutation (or by other reactions). The end products of hydrogen peroxide breakdown are oxygen and water. These enzymes lower the potential for the formation of the highly reactive hydroxyl radical as well as other energized, oxygen-containing species. Selenium is an essential component of glutathione peroxidase.

This enzyme is important in the decomposition of hydrogen peroxide and lipid peroxides (termination products of free radical attack on lipids). The nutritionally essential minerals are not antioxidants until they are incorporated into the protective antioxidant enzymes (Sies, 1985).

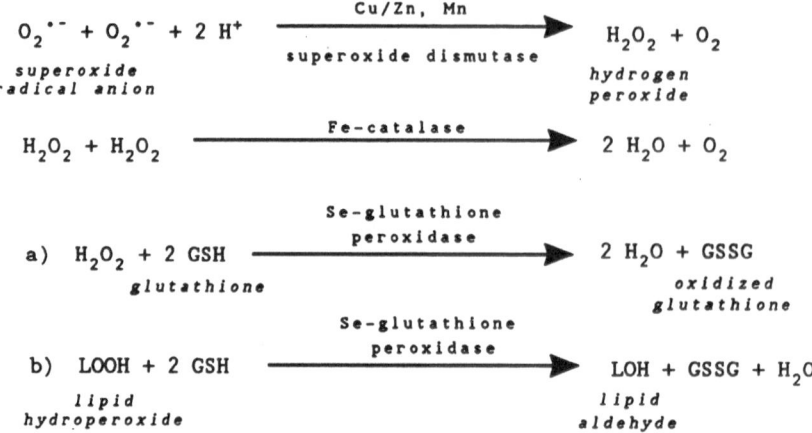

$$O_2^{\bullet-} + O_2^{\bullet-} + 2\,H^+ \xrightarrow[\text{superoxide dismutase}]{Cu/Zn,\ Mn} H_2O_2 + O_2$$

superoxide radical anion   hydrogen peroxide

$$H_2O_2 + H_2O_2 \xrightarrow{Fe\text{-catalase}} 2\,H_2O + O_2$$

a) $H_2O_2 + 2\,GSH \xrightarrow[\text{peroxidase}]{Se\text{-glutathione}} 2\,H_2O + GSSG$

glutathione   oxidized glutathione

b) $LOOH + 2\,GSH \xrightarrow[\text{peroxidase}]{Se\text{-glutathione}} LOH + GSSG + H_2O$

lipid hydroperoxide   lipid aldehyde

Fig. 3. Reactions involving the antioxidant enzymes.

Antioxidant Vitamins

Only three essential nutrients can directly interfere with the propagation stage of free radical generation and scavenge free radicals. Vitamin E (alpha tocopherol), the major lipid soluble antioxidant present in all cellular membranes, protects against lipid peroxidation (Machlin and Bendich, 1987). Vitamin C (ascorbic acid), is water soluble and along with vitamin E, can quench free radicals as well as singlet oxygen. Ascorbate can also regenerate the reduced, antioxidant form of vitamin E (Bendich et al., 1986). Recent work has shown that beta carotene, a pigment found in all photosynthetic plants, is an efficient quencher of singlet oxygen and can function as an antioxidant (Burton & Ingold, 1984). Beta carotene is the major carotenoid precursor of vitamin A. Vitamin A, however, cannot quench singlet oxygen and has a very small capacity to scavenge free radicals.

In addition to the direct action of these nutrients, riboflavin, a B vitamin, is a constituent of the enzyme glutathione reductase which is important in the regeneration of the antioxidant peptide, glutathione. Glutathione, in contrast to the other antioxidant defenses discussed, is a tripeptide composed of nonessential amino acids (Fig. 4.).

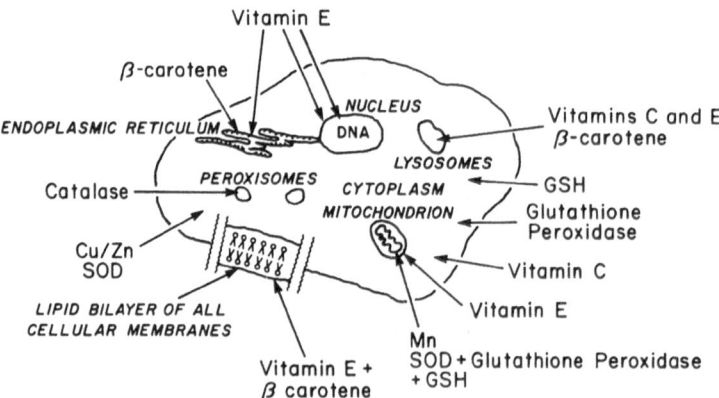

Fig. 4. Antioxidant protection within the cell.

It is important to note that the antioxidant enzymes are primarily intracellular and thus extracellular free radicals, either endogenously produced or from the environment, must be inactivated by the circulating antioxidants such as ceruloplasmin, a copper-containing protein and the direct acting vitamins discussed above (Machlin and Bendich, 1987).

Antioxidant Interactions

In addition to direct quenching of reactive, damaging free radicals, vitamin C has been clearly shown to interact with the tocopheroxyl radical and regenerate the reduced tocopherol. The evidence for this important antioxidant function for ascorbate includes cell-free experiments, investigations with liposomal membranes, and recently, *in vivo* evidence of higher concentrations of vitamin E in tissues of guinea pigs fed high dietary levels of vitamin C (Bendich et al., 1986).

Vitamin E can protect the conjugated double bonds of beta carotene from oxidation. The sparing action of tocopherol on beta carotene was first described *in vivo* in humans by Urbach et al. (1951).

Vitamin E can protect against many of the symptoms of selenium deficiency and vice versa (Combs, 1987). These sparing as well as synergistic actions are due to the ability of both tocopherol and selenium-dependent glutathione peroxidase to decrease the production of lipid peroxidation products (Fig. 5.).

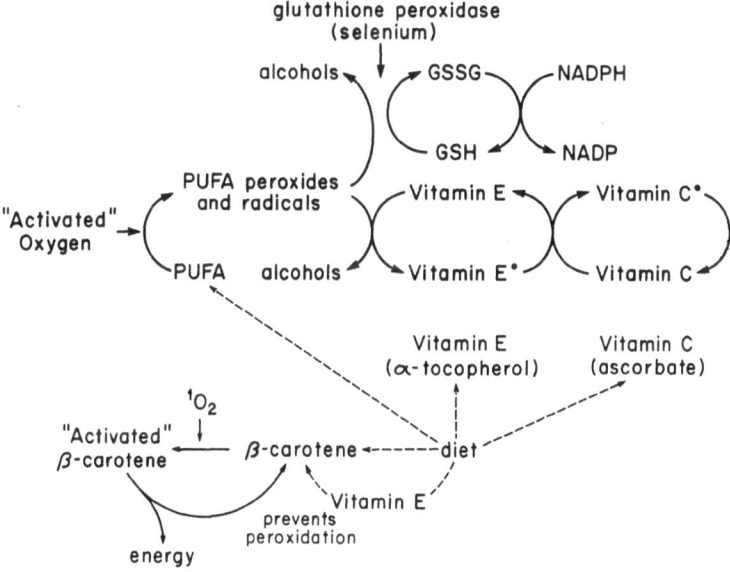

Fig. 5. Antioxidant interactions.

IMMUNE FUNCTIONS, FREE RADICAL REACTIONS AND ANTIOXIDANTS

Under normal circumstances, the magnitude of the antioxidant defenses available within the cell and extracellularly should be adequate to protect against oxidative damage. However, the balance can be lost because of overproduction of free radicals or by exposure to sources which overwhelm the antioxidant defenses.

Certain components in the diet can influence the level of free radicals generated in the body. High levels of dietary polyunsaturated fatty acids (PUFA) have been shown to be immunodepressing (Mertin and Hunt, 1976; Erickson et al., 1980; Newberne, 1981; Gurr, 1983). The unsaturated double bonds found in PUFA are prime targets for free radical damage and consequent initiation of chain reactions that result in the formation of lipid peroxides. Lipid

5

peroxides and aldehydes can alter cellular functions and even result in lysis of the oxidized cell membranes. Lipoproteins in the plasma can also be oxidized and these molecules are lymphotoxic (Cathcart et al., 1985).

The fluidity of cell membranes is, in part dependent upon the degree of unsaturation of its fatty acids. As the level of PUFA is increased, the potential for membrane lipid peroxidation is also increased. Lipid peroxidation causes a decrease in membrane fluidity (Meade and Mertin, 1978). In addition to consequent loss in fluidity, metabolites of lipid peroxidation can adversely affect immune responses (Mertin and Hughes, 1975; Erickson et al., 1983). Mice fed dietary lipid hydroperoxides showed marked atropy of the thymus as well as thymocyte necrosis (Oarada et al., 1988). Loss of membrane fluidity has been directly related to the decreased ability of lymphocytes to respond to challenges to the immune system (Fountain and Schultz, 1982).

The degree of unsaturation of the lipids incorporated into cell membranes can alter the level of exposure of membrane receptors, receptor number, and the potential for the synthesis of metabolites of arachidonic acid (Cinader et al., 1983; Wedner, 1984). Fatty acids derived from linoleic acid, the omega-6 fatty acids, are metabolized to more immunosuppressive products than those derived from eicosapentaenoic acid (omega-3, fish oils). Since prostaglandin and leukotriene syntheses involve the generation of free radicals, antioxidants can affect the concentrations of these critical immunomodulators. Dr. Fernandes discusses the effects of different fatty acids on immune responses in an autoimmune mouse model in chapter VIII(Fernandes, 1989).

LEUKOCYTE USE OF FREE RADICALS

Certain cells of the immune system use free radicals as weapons to destroy invading pathogens. Two major leukocytes that generate reactive oxygen intermediates are the neutrophils and the macrophages.

Neutrophils

The most abundant circulating phagocytic cells are called polymorphonucleated neutrophils. These cells respond to chemical signals released at sites of infection, and move towards the signal (chemotaxis). Neutrophils from vitamin C deficient guinea pigs have depressed chemotaxis and bactericidal activities (Shilotri, 1977). Vitamin C can enhance the chemotactic response of neutrophils from healthy individuals (Anderson, 1982; Beisel, 1982; Panush and Delafuente, 1985). Vitamin C an vitamin E supplementation, in separate studies, have also been found to normalize the reduced chemotactic and bactericidal activities of neutrophils from individuals with inherited phagocytosis disorders (Weening et al., 1981, Gallin, 1981; Boxer et al., 1979) as well as from newborn infants (Vohra et al., 1983). In several instances, clinical improvements were found. The depressed phagocytic function of neutrophils from premature infants was normalized following administration of vitamin E (Chirico et al., 1983). Neutrophils from healthy individuals given 1600 IU of vitamin E for two weeks showed enhanced chemotaxis and phagocytosis (Baehner et al., 1977).

Neutrophils have the capacity to take up molecular oxygen and generate reactive oxygen-containing molecules when stimulated. This is often called the oxidative burst. Free radicals and singlet oxygen, along with other reactive molecules, can kill bacterial pathogens directly or react with granulocyte-specific enzymes to form highly toxic compounds. Neutrophils can generate toxic halogenated molecules when the myeloperoxidase halide enzyme system is activated during the oxidative burst. The halogenated species can also lyse pathogens. The killing process is usually confined to intracellular vacuoles which enclose the phagocytized pathogen, however, these reactive molecules are also secreted and interact with nonphagocytized pathogens. During the oxidative burst, neutrophils take up vitamin C (Moser and Weber, 1983) and following activation, the vitamin C concentration is reduced (Hemila et al., 1985, Oberritter et al., 1986).

In addition to killing pathogenic organisms, these reactive molecules, when released into the surrounding area can cause mutations, lyse normal cells including other neutrophils, red blood cells and platelets, cause inflammation of surrounding tissues, inactivate protective enzymes and inhibit lymphocyte proliferation (Fantone and Ward, 1985). In addition, neutrophils can lyse tumor cells. Monocytes seem to protect lymphocytes from neutrophil oxidative injury because monocytes have high levels of catalase. Bacteria containing high levels of superoxide dismutase or carotenoids are more resistant to oxidative damage than bacteria lacking this antioxidant enzyme or singlet oxygen quencher (Clark and Nauseef, 1988).

Anderson et al. (1987) have recently shown that vitamin C can decrease the damaging effects of the products of the oxidative burst without decreasing the intracellular concentration of reactive, bactericidal molecules. In addition, vitamin C protects alpha 1 protease inhibitor (a protective enzyme found in the lung) from inactivation by the free radicals generated during the oxidative burst (Theron and Anderson, 1985).

Neutrophils contain superoxide dismutase, catalase and glutathione/ glutathione peroxidase to protect against autooxidation but are still damaged by the reactive oxygen species. Investigators have examined the role of selenium in bactericidal and cytotoxic activities involving oxidative mechanisms (Spallholz, 1981a & b). However, there is evidence that glutathione peroxidase deficiency (as the result of dietary selenium deficiency) does not decrease neutrophil bactericidal activity (Bass et al., 1977). The susceptibility of selenium-deficient neutrophils to alterations in the respiratory burst is examined in chapter XII written by Dr. Spallholz (Spallholz et al., 1989). Fraker and Wirth (1989) have shown that zinc deficiency decreases the ability of murine phagocytic cells to kill intracellular parasitic pathogens. The mechanism by which zinc enhances oxygen radical production in this model is examined in chapter X. Copper deficiency also adversely affects neutrophil functions. Dr. Prohaska reviews these data in chapter XI (Prohaska, 1989).

Vitamin E-deficient rats have impaired neutrophil and macrophage chemotaxis, reduced ingestion of complement coated beads and decreased protection from autooxidative damage (Harris et al., 1980). Neutrophil-induced oxidative damage to lung tissue following burn injury was significantly reduced when rats were preinjected with vitamin E (Till et al. 1985). In an extension of earlier investigations, Dr. Boxer presents new techniques for the evaluation of the critical role of oxidative reactions and antioxidants, such as vitamin E, in the functioning of neutrophils in chapter III (Boxer, 1989).

Macrophages

Whereas neutrophils are mainly associated with acute inflammatory responses, macrophages are involved with chronic inflammation. Macrophages, unlike neutrophils, are central to the development of specific immune responses and process antigens for presentation to lymphocytes. Both these phagocytic cells have the cell-membrane associated NADPH-oxidase system thereby permitting macrophages to synthesize and secrete reactive oxygen species including hydrogen peroxide, hydroxyl radicals and superoxide radicals. In addition, prostaglandins, leukotrienes, interleukin 1 and interferons are produced by macrophages. Interleukin 1 stimulates several cell types, cocultured with macrophages, to synthesize superoxide dismutase (Masuda et al., 1988). The consequences of this enzyme induction are being explored.

Chronic immune-mediated inflammation, such as found in experimentally induced granulomas, has been decreased in animals given superoxide dismutase, catalase or vitamin E, suggesting a free radical-associated etiology (Chensue et al., 1984). Macrophage-generated oxygen radicals have been shown to enhance kidney damage in animal models of glomerulonephritis. Treatment with polyethylene glycol-modified catalase decreased this damage (Johnson, 1986).

The synovial fluid of the joints of rheumatoid arthritis patients contains high levels of reactive oxygen species. Infiltrates of both neutrophils and T lymphocytes are found in the affected joints. Local lipid peroxidation has been correlated with the degree of inflammation in animal models of arthritis. Increased production of autoimmune immunoglobulins (rheumatoid factor) indicating a perturbation in B cell function is also found in this disease. Clinical intervention studies have not been initiated with ascorbic acid, vitamin E or beta carotene, however, this may be efficacious since administration of antioxidants such as superoxide dismutase and catalase into arthritic joints decreased inflammation (Sies, 1985). Ascorbic acid levels are low in patients with rheumatoid arthritis and cannot be explained by decreased intake or absorption (Lunec and Blake, 1988).

Human milk contains high levels of the antioxidant micronutrients discussed in this symposium. Measurable levels of superoxide dismutase, catalase and glutathione peroxidase as well as higher than maternal plasma levels of vitamins C and E and beta carotene help protect the newborn digestive system from damaging inflammatory responses. Dr. Goldman discusses the antiinfective and antioxidant nature of human milk in chapter VI (Goldman, 1989).

FREE RADICALS AND LYMPHOCYTE FUNCTIONS

The major leukocytes involved in the generation of specific immune responses are the T and B lymphocytes. The vitamin C and E content of lymphocytes and mononuclear cells is usually an order of magnitude greater than that found in platelets and red blood cells (Moser and Bendich, 1989; Hatam and Kayden, 1979). If splenocytes from mice fed standard diets are exposed *in vitro* to substances which induce peroxidative damage through the generation of free radicals, both T and B lymphocyte proliferative responses to mitogens and *in vitro* antigen induced antibody formation by B lymphocytes are inhibited. When additional vitamin E or other antioxidants are added to these cultures, the immunosuppression is overcome (Hoffeld, 1981, 1983).

Single essential nutrient deficiencies adversely affect many aspects of T and B lymphocyte function. In the clinical setting, a single deficiency is rare. Most often the nutritional problems involve both macro and micronutrients. Dr. Chandra examines the clinical effects of deficiencies, emphasizing the role of antioxidant nutrients in chapter II (Chandra, 1989). In the field of veterinary medicine, single and multiple micronutrient deficiencies have been associated with increased rates of infection in farm production animals. Drs. Tengerdy and Frye review the effects of deficiencies as well as supplementation (in particular with vitamin E) on the disease resistance of sheep, cattle, chickens and pigs in chapter IX (Tengerdy, 1989).

Laboratory animals, fed well defined diets, have been studied to determine the effects of copper and selenium on specific immune responses. Deficiency of either essential nutrient affects T lymphocytes to a greater degree than B lymphocytes. These data are discussed by Dr. Prohaska (1989) (chapter XI) and Dr. Spallholz (1989) (chapter XII). The immunotoxic effects of high levels of selenium are also reviewed.

The immunoenhancing effects of supplemental vitamin E and vitamin C and beta carotene are discussed in chapter IV (Bendich, 1989). In laboratory animals, T and B cell proliferative responses have correlated with dietary and serum vitamin E levels over a thousand-fold increase in dietary vitamin E (from deficient to 1000 mg/kg diet; Bendich et al., 1986).

As laboratory animals age, T lymphocyte functions are reduced to a greater degree than B lymphocyte activities. T lymphocyte membranes are more fluid than B cell membranes in young mice. However, as mice age, T cells lose their fluidity whereas B cells retain the same level. T cell lipids are more susceptible to peroxidation than B cell lipids (Hendricks and Heidrick, 1988). The ability of T lymphocytes to form rosettes was significantly inhibited following exposure to oxygen radicals whereas B lymphocyte rosette formation was not significantly affected (Grever et al., 1980).

As the aging process progresses, the level and activities of the antioxidant enzymes decrease. Supplementation with the minerals required for enzyme activity may not increase protection because the defect may be in the synthesis of the protein portion of the enzyme. Even though the selenium-containing glutathione peroxidase, can lower prostaglandin levels, the activity of this enzyme is decreased in the elderly compared to young adults.

In contrast, supplementation with the antioxidant vitamins may lower the level of peroxidation and maintain immune responses. This hypothesis has recently been tested in elderly subjects and is reported by Dr. Meydani in chapter V (Meydani et al., 1989). She had earlier found that elderly mice given vitamin E had *in vivo* and *in vitro* immune responses restored to the levels found in young mice. Concurrently, the production of immunosuppressive prostaglandins was reduced in the elderly mice and humans (Meydani et al., 1984). Since the formation of prostaglandins requires free radical generation, vitamin E can alter the levels of prostaglandins formed. The findings of immunosuppression associated with cellular membrane peroxidation may help to explain the beneficial effects of vitamin E on the immune responses.

T lymphocytes have several functions including helper and cytotoxic activities. Cytotoxic T cells are involved in the recognition and killing of tumor cells. Recent studies have shown that dietary beta carotene and carotenoids of similar chemical structure (but lacking provitamin A activity) enhanced cytotoxic T cell activity and lowered tumor levels in animal models (Bendich, 1988). Dr. Schwartz discusses his data in chapter VII (Schwartz and Shklar, 1989) and Dr. Bendich reviews the immunoenhancing effects of carotenoids in chapter IV (Bendich, 1989).

SUMMARY

This short introduction encompasses only a small portion of the literature linking free radical production and consequent effects on immune functions. The role of essential dietary components in modulating these effects is an area of intense and expanding investigation. Each of the nutrients examined in the following chapters has distinct functions, but in this volume we concentrate on their shared capacity to act as antioxidants.

REFERENCES

Anderson, R., 1982, Ascorbic acid and immune functions: Mechanism of immunimmunostimulation, in: "**Vitamin C Ascorbic Acid**," J. N. Counsell, D. H. Hornig, eds., Applied Science, London.

Anderson, R., Lukey, P. T., Theron, A. J., and Dippenaar, U., 1987, Ascorbate and cysteine-mediated selective neutralisation of extracellular oxidants during N-formyl peptide activation of human phagocytes, **Agents and Actions**, 20(1/2):77.

Baehner, R. L., Boxer, L. A., Allen, J. M., and Davis, J., 1977, Autoxidation as a basis for altered function by polymorphonuclear leukocytes, **Blood**, 50:327.

Bass, D. A., DeChatelet, L. A., Burk, R. F., Shirley, P., and Szejda, P., 1977, Polymorphonuclear leukocyte bactericidalactivity and oxidative metabolism during glutathioneperoxidase deficiency, **Infect. Immun.**, 18:78.

Beisel, W. R., 1982, Single nutrients and immunity, **Amer. J. Clin. Nutr.**, 35:417.

Bendich, A., 1988, A role for carotenoids in immune function, **Clinical Nutrition** 7(3):113.

Bendich, A., 1989, Effects of antioxidant vitamins on cellular immune functions, in: "**Antioxidant Nutrients and Immune Functions**," A. Bendich, M. Phillips, A. Tengerdy, eds., Pergamon Press, New York.

Bendich, A., Gabriel, E., and Machlin, L. J., 1986, Dietary vitamin E requirement for optimum immune responses in the rat, **J. Nutr.**, 116(4):675.

Bendich, A., Machlin, L. J., Scandurra, O., Burton, G. W., and Wayner, D. D. M., 1986, The antioxidant role of vitamin C, **Adv. in Free Radical Biol. & Med.**, 2:419.

Boxer, L. A., 1989, Functional effects of leukocyte antioxidants on polymorphnuclear leukocyte (PMN) behavior, in: "**Antioxidant Nutrients and Immune Functions**," A. Bendich, M. Phillips, R. Tengerdy, eds., Pergamon Press, New York.

Boxer, L. A., Oliver, J. M., Spielberg, S. P., Allen, J. M., and Schulman.J.D.,1979, Protection of granulocytes by vitamin E in glutathione synthetase deficiency, **New Eng. J. Med.**, 301:901.

Burton, G. W. and Ingold, K. U., 1984, Beta-Carotene: an unusual type of lipid antioxidant, **Science**, 224:569.

Cathcart, M. K., Morel, D. W., and Cisolm, G. M. I., 1985, Monocytes and neutrophils oxidize low density lipoprotein making it cytotoxic, **J. Leukocyte Biol.**, 38:341.

Chandra, R. K., 1989, Molecular and cellular basis of nutrition-immunity interactions, in: "**Antioxidant Nutrients and Immune Functions**," A. Bendich, M. Phillips, R.Tengerdy, eds., Pergamon Press, New York.

Chensue, S. W., Quinlan, L., Higashi, G. I., and Kunkel, S. L., 1984, Role of oxygen reactive species in Schistosoma mansoni egg-induced granulomatous inflammation, **Biochem. Biophys. Res. Commun.**, 122:184.

Chirico, G., Marconi, M., Colombo, A., Chiara, A., Rondini, G., and Ugazio, G., 1983, Deficiency of neutrophil phagocytosis in premature infants: Effect of vitamin E supplementation, **Acta Paediatr. Scand.**, 72:521.

Cinader, B., Clandinin, M. T., Hosokawa, T., and Robblee, N. M., 1983, Dietary fat alters the fatty acid composition of lymphocyte membranes and the rate at which suppressor capacity is lost, **Immunol. Letters**, 6:331.

Clark, R. A. and Nauseef, W. M., 1988, Phagocyte oxidants and protective mechanisms *in*: **"Cellular Antioxidant Defense Mechanisms, Volume I,"** Ching Kuang Chow, ed., CRC Press, Inc., Boca Raton, Florida.

Combs, G. F. J., 1987, Protective roles of minerals against free radical tissue damage, AIN Symposium Proceedings, **Nutrition '87**:55.

Cutler, R. G., 1986, Aging and oxygen radicals, *in*: **"Physiology of oxygen radicals,"** A. E. Taylor, S. Matalon, P. A. Ward, eds., The Williams & Wilkins Company, Baltimore, Maryland.

Dormandy, T. L., 1983, An approach to free radicals, **Lancet**, 8357:1010.

Erickson, K. L., Adams, D. A., and McNeill C.J., 1983, Dietary lipid modulation of immune responsiveness, **Lipids**, 18:468.

Erickson, K. L., McNeill, C. J., Gershwin M.E, and Ossmann, J. B., 1980, Influence of dietary fat concentration and saturation on immune ontogeny in mice, **J. Nutr.**, 110:1555.

Fantone, J. C. and Ward, P. A., 1985, Polymorphonuclear leukocyte-mediated cell and tissue injury, **Hum. Pathol.**, 16(10):973.

Fernandes, G., 1989, Effect of Omega-3 fatty acids and vitamin -E supplements on autoimmune disease, *in*: **"Antioxidant Nutrients and Immune Functions,"** A. Bendich, M. Phillips, R. Tengerdy, eds., Pergamon Press, New York.

Flohe, L., Beckmann, R., Giertz, H., and Loschen, G., 1985, Oxygen-centered free radicals as mediators of inflammation, *in*: **"Oxidative Stress,"** H. Sies, ed., Academic Press, London.

Fountain, M. W. and Schultz, R. D., 1982, Effects of enrichment of phosphatidylcholine liposomes with cholesterol or alpha-tocopherol on the response of lymphocytes to phytohemagglutinin, **Molecular Immunology**, 19:59.

Fraker, P. J. and Wirth, J. J., 1989, Oxygen metabolites of phagocytic cells altered by zinc deficiency, *in*: **"Antioxidant Nutrients and Immune Functions,"** A. Bendich, M. Phillips, R. Tengerdy, eds., Pergamon Press, New York.

Freeman, B. A. and Crapo, J. D., 1982, Biology of disease: Free radicals and tissue injury, **Laboratory Investigation**, 47(5):412.

Gallin, J. I., 1981, Abnormal phagocyte chemotaxis: Pathophysiology, clinical manifestations, and management of patients, **Rev. Infect. Dis.**, 3(6):1196.

Gey, F., 1986, On the antioxidant hypothesis with regard to arteriosclerosis, **Biblthca. Nutr. Dieta.**, 37:53.

Gey, K. F., Brubacher, G. B., and Stahelin, H. B., 1987, Plasma levels of antioxidant vitamins in relation to ischemic heart disease and cancer, **Am. J. Clin. Nutr.**, 45:1368.

Goldman, A. S., 1989, Antioxidant and other anti-inflammatory properties of human milk, *in*: **"Antioxidant Nutrients and Immune Functions,"** A. Bendich, M. Phillips, R. Tengerdy, eds., Pergamon Press, New York.

Grever, M. R., Thompson, V. N., Balcerzak, S. P., and Sagone, A. L., 1980, The effect of oxidant stress on human lymphocyte cytotoxicity, **Blood**, 56:284.

Grisham, M. B. and McCord, J. M., 1986, Chemistry and cytotoxicity of reactive oxygen metabolites, *in*: **"Physiology of Oxygen Radicals,"** A. E. Taylor, S. Matalon, P. A. Ward, eds., The Williams & Wilkins Company, Baltimore, Maryland.

Gurr, M. I., 1983, The role of lipids in the regulation of the immune system, **Prog. Lipid Res.**, 22:257.

Halliwell, B. and Gutteridge, J. M. C., 1985, Lipid peroxidation: a radical chain reaction, *in*: **"Free Radicals in Biology and Medicine,"** B. Halliwell, J. M. C. Gutteridge, eds., Clarend Press, Oxford.

Harris, R. E., Boxer, L. A., and Baehner, R. L., 1980, Consequences of vitamin E deficiency on the phagocyte and oxidative function of the rat polymorphonuclear leukocyte, **Blood**, 55:338.

Hatam, L. J. and Kayden, H. J., 1979, A high-performance liquid chromatographic method for the determination of tocopherol in plasma and cellular elements of the blood, **J. Lipid Res.**, 20:639.

Hemila, H., Roberts, P., and Wikstrom, M., 1985, Activated polymorphonuclear leucocytes consume vitamin C, **Febs Lett.**, 178:25.

Hendricks, L. C. and Heidrick, M. L., 1988, Susceptibility to lipid peroxidation and accumulation of fluorescent products with age is greater in T-cells than B-cells, **Free Radical Biology & Medicine**, 5:145.

Hirschelmann, R. and Bekemeier, H., 1981, Effects of catalase, peroxidase, superoxide dismutase and 10 scavengers of oxygen radicals in carrageenin oedema and in adjuvant arthritis of rats, **Experientia**, 37:1313.

Hoffeld, J. T., 1981, Agents which block membrane lipid peroxidation enhance mouse spleen cell immune activities *in vitro*: Relationship to enhancing activity of 2-mercaptoethanol, **Eur. J. Immunol.**, 11:371.

Hoffeld, J. T., 1983, Inhibition of lymphocyte proliferation and antibody production *in vitro* by silica, talc, bentonite or corynebacterium parvum: Involvement of peroxidative processes, **Eur. J. Immunol.**, 13:364.

Johnson, K. J., 1986, Neutrophil-independent oxygen radical-mediated tissue injury, *in*: **"Physiology of Oxygen Radicals,"** A. E. Taylor, S. Matalon, P. Ward, eds., The Williams & Wilkins Company, Baltimore, Maryland.

Krinsky, N. I. and Deneke, S. M., 1982, Interaction of Oxygen and Oxy-radicals With Carotenoids, **JNCI**, 69(1):205.

Lunec, J. and Blake, D., 1988, Oxidative damage and its relevance to inflammatory joint disease, *in*: **"Cellular Antioxidant Defense Mechanisms,"** Ching Kuang Chow, ed., CRC Press, Inc., Boca Raton, Florida.

Machlin, L. J., 1984, Vitamin E, *in*: **"Handbook of Vitamins,"** L. J. Machlin, ed., Marcel Dekker, Inc., New York.

Machlin, L. J. and Bendich, A., 1987, Free radical tissue damage: protective role of antioxidant nutrients, **FASEB J.**, 1:441.

Masuda, A., Longo, D. L., Kobayashi, Y., Appella, E., Oppenheim, J. J., and Matsushima, K., 1988, Induction of mitochondrial manganese superoxide dismutase by interleukin 1, **FASEB J.**, 2(15):3087.

Meade, C. J. and Mertin, J., 1978, Fatty acids and immunity, **Adv. Lipid Res.**, 16:127.

Mertin, J. and Hughes, D., 1975, Specific inhibitory action of polyunsaturated fatty acids on lymphocyte transformation induced by PHA and PPD, **Int. Archs. Allergy Appl. Immunol.**, 48:203.

Mertin, J. and Hunt, R., 1976, Influence of polyunsaturated fatty acids on survival of skin allografts and tumor incidence in mice, **Proc. Nat. Acad. Sci. USA**, 73:928.

Meydani, S. N., Blumberg, J. B., Yogeeswaran, G., and Meydany, M., 1989, Antioxidants and the aging immune system, *in*: **"Antioxidant Nutrients and Immune Functions,"** A. Bendich, M. Phillips, R. Tengerdy, eds., Pergamon Press, New York.

Meydani, S. N., Meydani, M., Verdon, C. P., Blumberg, J. B., Hayes, C., 1984, PGE2 control of vitamin E-enhanced immunity in old mice, **Fed. Proc.**, 43:478.

Moser, U. and Bendich, A., 1989, Vitamin C, *in*: **"Handbook of Vitamins - Volume 2,"** in press.

Moser, U. and Weber, F., 1984, Uptake of ascorbic acid by human granulocytes, **Internat. J. Vit. Nutr. Res.**, 54:47.

Newberne, P. M., 1981, Dietary fat, immunological response, and cancer in rats, **Cancer Res.**, 41:3783.

Oarada, M., Ito, E., Trao, K., Miyazawa, T., Fujimoto, K., and Kaneda, T., 1988, The effect of dietary lipid hydroperoxide on lymphoid tissues in mice, **Biochimica et Biophysica Acta**, 960:229.

Oberritter, H., Glatthaar, B., Moser, U., and Schmidt, K. H., 1986, Effect of functional stimulation on ascorbate content in phagocytes under physiological and pathological conditions, **Int. Archs. Allergy Appl. Immun.**, 81:46.

Panush, R. S. and Delafuente, J. C., 1985, Vitamins and immunocompetence, **Wld. Rev. Nutr. Diet**, 45:97.

Prohaska, J. R., 1989, Copper deficiency alters the immune system, *in*: **"Antioxidant Nutrients and Immune Functions,"** A. Bendich, M. Phillips, R. Tengerdy, eds., Pergamon Press, New York.

Schwartz, J. L. and Shklar, G., 1989, Prevention and regression of hamster oral squamous cell carcinoma following administration of carotenoids, *in*: **"Antioxidant Nutrients and Immune Functions,"** A. Bendich, M. Phillips, R. Tengerdy, eds., Pergamon Press, New York.

Shilotri, P. G., 1977, Glycolytic, hexose monophosphate shunt and bactericidal activities of leukocytes in ascorbic acid deficient guinea pigs, **J. Nutr.**, 107:1507.

Sies, H., 1985, **"Oxidative Stress,"** Academic Press, London.

Slater, T. F., 1972, Free radical mechanisms in tissue injury, Pion Limited, London.

Slater, T. F., 1987, Free radical-mediated tissue damage, AIN Symposium Proceedings. **Nutrition '87**:46.

Spallholtz, J. E., 1981, Selenium: What role in immunity and immune cytotoxity? *in*: **"Selenium in Biology and Medicine,"** J. E. Spallholtz, J. L. Martin, H. E. Ganther, eds., Avi Publishing Co., Connecticut.

Spallholz, J. E., 1981, Anti-inflammatory, immunologic and carcinostatic attributes of selenium in experimental animals, *in*: **"Diet and Resistance to Disease,"** M. Phillips, A. Baetz, eds., Plenum Press, New York.

Spallholz, J. E., Boylan, L. M., and Larsen, H. S., 1989, Selenium: antioxidant and nutrient component of the immune system, *in*: **"Antioxidant Nutrients and Immune Functions,"** A. Bendich, M. Phillips, R. Tengerdy, eds., Pergamon Press, New York.

Tengerdy, R. and Frye, T. M., 1989, Feeding increased levels of vitamin E for immunity and disease resistance, *in*: **"Antioxidant Nutrients and Immune Functions,"** A. Bendich, M. Phillips, R. Tengerdy, eds., Pergamon Press, New York.

Theron, A. and Anderson, R., 1985, Investigation of the protective effects of the antioxidants ascorbate, cysteine, and dapsone on the phagocyte-mediated oxidative inactivation of human alpha-1-protease inhibitor *in vitro*, **Am. Rev. Respir. Dis.**, 132:1049.

Till, G. O., Hatherill, J. R., Tourtellote, W. W., Lutz, M. J., and Ward, P. A., 1985, Lipid peroxidation and acute lung injury after thermal trauma to skin, **Am. J. Pathol.**, 119:376.

Urbach, C., Hickman, K., and Harris, P. L., 1951, Effect of individual vitamins A, C, E and carotene administered at high levels and their concentration in the blood, **Exp. Med. Surg.**, 10:7.

Vohra, K., Khan, A. J., Rosenfeld, W., Telang, V., and Evans, H. E., 1983, Correction of defective chemotaxis of neonatal neutrophils with ascorbic acid, **Pediatr. Res.**, 17:340.

Wedner, H. J., 1984, Biochemical events associated with lymphocyte activation, **Surv. Immunol. Res.**, 3:295.

Weening, R. S., Schoorel, E. P., Roos, D., van Schaik, M. L., Voetman, A., Bot, A. A., Batenburg-Plenter, A. M., Willems, C., Zeijlemaker, W. P., Astaldi, A., 1981, Effect of ascorbate on abnormal neutrophil, platelet and lymphocyte function in a patient with Chediak-Higashi syndrome, **Blood**, 57:856.

CELLULAR AND MOLECULAR BASIS OF NUTRITION-IMMUNITY INTERACTIONS

Ranjit Kumar Chandra

Departments of Pediatrics and Medicine
Health Sciences Centre
Memorial University of Newfoundland
St. John's, Newfoundland A1B 3V6, Canada

INTRODUCTION

It is now established that nutrition is a critical determinant of immunocompetence and risk of illness. Young children with protein-energy malnutrition exhibit increased mortality and morbidity, due largely to infectious disease. Recent work has demonstrated that undernourished individuals have impaired immune responses. The most consistent abnormalities are seen in cell-mediated immunity, complement system, phagocytes, mucosal secretory antibody response, and antibody affinity. These changes, together with other handicapping factors observed in underprivileged societies, lead to more infections, which in turn produce physiological changes that worsen nutritional status. It is now established that deficiencies of single nutrients also impair immune responses. The best studied are zinc, iron, vitamin B6, vitamin A, copper, and selenium. If malnutrition occurs during fetal life, as epitomized by small-for-gestational age infants, the effects on cell-mediated immunity are very significant and long lasting. There is much recent evidence to suggest that at the other end of the age spectrum, namely in old age, nutrition plays an important role in maintenance of optimum immunity. These interactions of nutrition and immunity have several practical applications. In this interpretative review, a summary of current findings is given.

BACKGROUND AND EPIDEMIOLOGY

Malnutrition is the most frequent cause of immunodeficiency worldwide. The common occurrence of nutritional deficiencies in developing countries is well known. It has now been revealed that mild to moderate undernutrition is not uncommon even in industrialized countries. For example, iron deficiency is seen in at least 9 percent of North American population. In hospitalized patients, obvious malnutrition is frequent.

The combination of malnutrition and infection is responsible for much of the morbidity and mortality in all age groups. In patients with a variety of primary diseases, such as cancer and Crohn's inflammatory bowel disease, nutritional deficiencies further complicate the picture and increase the risk of infectious complications. Although much of the initial work on nutrition and immunity was done on young children in developing countries, the general principle that nutrition is a critical determinant of immunocompetence is applicable universally.

The initial studies in this field were done in Asia and Africa. "To convey a sense of time, the discipline of immunology was not then even exalted by the general use of terms such as "cell-mediated immunity," "T-lymphocyte subsets," "immunoregulation," and so on. In malnourished patients, we found impaired delayed cutaneous hypersensitivity, lymphocyte proliferation response to mitogens, complement activity, and secondary antibody response to certain antigens. These findings were soon confirmed by several investigators. Our subsequent work has demonstrated that protein-energy malnutrition results in a reduced number of rosetting T-lymphocytes, increased deoxynucleotidytl transferase activity, decreased serum thymic factor, fewer helper T-cells, impaired production of gamma-interferon and interleukin-2, reduced antibody affinity, impaired secretory IgA antibody response, and phagocyte dysfunction. Malnutrition, however, is not a single entity but rather a broad syndrome. We now know that deficiencies of trace elements and vitamins impair immunity. Both in humans and laboratory animals, intrauterine malnutrition causes prolonged, even permanent, depression of immunity in the offspring. Furthermore, nutrition is an important determinant of waning immunity in old age" (Citation Classic 1987).

Epidemiological studies have documented the adverse effect of protein-energy malnutrition (PEM) on morbidity and mortality. Pathological examination of tissues from children dying of PEM showed the frequent presence of several opportunistic microorganisms including Pneumocystis carinii. Morbidity due to diarrhoeal disease is increased particularly among those children whose weight-for-height is less than 70 percent of standard.

Lymphoid tissues show a significant atrophy. For instance, the size of the thymus is small. Histologically, there is a loss of corticomedullary differentiation, there are fewer lymphoid cells, and the Hassal bodies are enlarged, degenerated, and occasionally, calcified. In the spleen, there is a loss of lymphoid cells around small blood vessels. In the lymph node, the thymus-dependent areas show depletion of lymphoid cells.

IMMUNE RESPONSES IN PROTEIN-ENERGY MALNUTRITION

This topic has been the subject of several reviews (Chandra and Newberne, 1977; Beisel, 1982; Chandra, 1983, 1986, 1988; Keusch et al. 1983; McMurray 1984). Delayed cutaneous hypersensitivity responses both to recall and new antigens are markedly depressed. There is a significant positive correlation between the size of skin response and visceral protein synthesis as judged by serum albumin concentration. It is not uncommon to have complete anergy to a battery of different antigens. These changes are observed in moderate deficiencies as well. Findings in patients with kwashiorkor were more striking compared with those in marasmus. One plausible reason for reduced cell-mediated immunity in PEM is the reduction in mature fully differentiated T lymphocytes that can be recognized by the classical technique of rosette-formation or by the newer method of fluorescent labelling with monoclonal antibodies. The reduction in serum thymic factor activity observed in PEM may underlie the impaired maturation of T lymphocytes. There is an increase in the amount of deoxynucleotidyl transferase activity in leukocytes, a feature of immaturity. The recent availability of monoclonal antibodies has provided an excellent tool for the identification and enumeration of subsets of T cells. Cell flow methods showed that the number of helper T4 cells was decreased markedly, often to values less than 50% of controls. The change in number of suppressor T cells is less marked. Thus the helper/suppressor ratio is significantly decreased. Lymphocyte proliferation and synthesis of DNA are reduced, especially when autologous patient plasma is used in cell cultures. This may be the result of inhibitory factors as well as deficiency of essential nutrients lacking in patient's plasma.

Serum antibody responses are generally intact in PEM, particularly when antigens in adjuvant are administered or in the case of those materials that do not evoke T cell response. Rarely, the antibody response to organisms such as Salmonella typhi may be decreased. Before impaired antibody response can be attributed to nutritional deficiency, one must carefully rule out infection as a confounding factor. Recently, we have found that antibody affinity is decreased in patients who are malnourished. This may provide an explanation for a higher frequency of antigen-antibody complexes found in such patients. As opposed to serum antibody responses, secretory IgA antibody levels after deliberate immunization with viral vaccines are decreased; there is a selective reduction in secretory IgA levels. This may have several clinical implications, including an increased frequency of septicemia in undernourished children.

There are several distinct steps in phagocytosis and destruction of microorganisms, including production and mobilization of phagocytes, opsonization, ingestion, metabolic activity, killing and extrusion. The process of phagocytosis is also affected in PEM. Complement is an essential opsonin and the levels and activity of most complement components are decreased. The best documented is a reduction in complement C3, factor B, and total hemolytic activity. Although the ingestion of particles by phagocytes is intact, subsequent metabolic activation and destruction of bacteria is reduced.

There is much recent interest in the role of cytokines in promoting and amplifying immune responses. A number of distinct substances have been identified and their physiological roles studied. These are polypeptide compounds with a wide range of molecular weight. Included in the human lymphokines are interleukin-1, interleukin-2, interleukin-3, colony stimulating factors, interleukin-4, B-cell differentiating factor, interferons (alpha, beta, gamma) and tumour necrosis factor (Dinarello and Mier, 1987). Changes in the production and function of various lymphokines and monokines in PEM have been reviewed recently (Hoffman-Goetz, 1988). Briefly, macrophage migration inhibition factor has been found to be decreased whereas there is little change in leukocyte migration inhibition factor. Gamma interferon production is decreased. Interleukin-1 production and functional activity on target cells is depressed. Interleukin-2 production may be normal but there are alterations in IL-2 receptor binding.

SELECTED NUTRIENT DEFICIENCIES

Considerable work in animals deprived of one dietary element and findings in rare patients with a single nutrient deficiency have confirmed the crucial role of several vitamins and trace elements in immuncompetence. For example, deficiencies of pyridoxine, folic acid, vitamin A, vitamin C and vitamin E result in impaired cell-mediated immunity and reduced antibody responses. Vitamin B6 deficiency results in decreased lymphocyte stimulation response to mitogens such as phytohemagglutinin. A moderate increase in vitamin A intake enhances immune response and affords partial protection against the development of certain tumors in animals (Watson and Rybski, 1988). Similarly, moderate doses of beta-carotene increase the number of CD4+ cells in human volunteers and there is a slight increase in lymphocye stimulation response to mitogens (Chandra, Hambreaus, and Puri, 1989, Unpublished data). Moderate excess of vitamin E and selenium enhances immunocompetence and increases resistance to microorganisms (Bendich, 1988).

Zinc deficiency, both acquired and inherited, is associated with lymphoid atrophy, decreased cutaneous delayed hypersensitivity responses and homograft rejection, and lower thymic hormone activity. In laboratory animal models these findings can be confirmed and in addition one can demonstrate reduced number of antibody-forming cells in the spleen and impaired T-killer cell activity. Wound healing is impaired. Excess zinc also depresses neutrophil function and lymphocyte responses. Deficiency of iron is the commonest nutritional problem worldwide, even in industrialized countries. On the one hand, free iron is necessary for bacterial growth : removal of iron with the help of lactoferrin or other chelating agents reduced bacterial multiplication, particularly in the presence of specific antibody. On the other hand, iron is needed by neutrophils and lymphocytes for optimal function : bactericidal capacity is reduced in iron deficiency. Also, the lymphocyte proliferation response to mitogens and antigens is impaired : response to tetanus toxoid and herpes simplex antigens was low in iron-deficiency subjects and iron therapy resulted in a significant improvement in their response. There are many molecular explanations for impaired lymphocyte and neutrophil function in iron deficiency, including the deficiency of myeloperoxidase and ribonucleotidyl reductase. Copper-deficient animals show a reduction in the number of antibody-producing cells compared to healthy and pair-fed controls. Thymic factor activity is reduced.

The effect of single nutrient deficiency on serum thymic hormone activity has been evaluated recently. Zinc is critical to the biological activity of thymic inductive factors; as much as 80% of such activity is lost when zinc is chelated. Vitamin B6 also exerts a significant influence on thymic factor activity. On the other hand, the activity of thymulin is not affected significantly by deficiencies of vitamin A and selenium.

YOUNG INFANTS AND THE ELDERLY

At first glance, a discussion of immunological changes brought about by nutritional deficiencies in young infants and in the elderly appears to be an illogical combination. However, there are many striking similarities between the two ends of the age spectrum. Both neonates and elderly have suboptimal immune responses and are susceptible to infection. When nutritional deficiency complicates the picture, impairment of immunocompetence is more marked and longer-lasting.

The immune system develops during pregnancy and the first few months after birth. If the infant is born prematurely or if he exhibits growth retardation as a result of a number of environmental factors, including maternal malnutrition or infection, immunocompetence is reduced. The impact on T-lymphocyte numbers and cell-mediated immunity is most discernible. The preterm low-birth-weight infant generally recovers its ability to mount immune response by the age of 3 months. However, the small-for-gestational-age infant may continue to show reduced cell-mediated immunity for several months and years. There is a significant difference in the immunocompetence of SGA infants who exhibit higher morbidity rate and that of infants with lower morbidity. In laboratory animal models of intrauterine malnutrition, immune responses are impaired both in first- and second-generation offspring (Chandra, 1975).

The nutritional regulation of immunity and risk of infection in old age has been reviewed recently (Chandra, 1989). Many elderly individuals show a progressive loss of immune function. Cell-mediated immunity is impaired. This may be due in part to marked reduction in the putative thymic hormone(s). It is interesting that about 35 percent of subjects above 65 years retain immunocompetence at levels seen in young adult life. Other studies have documented alterations in nutritional status and body composition, including decreased lean body mass, loss of visceral protein, and increase in the relative proportion of body fat. These changes in body constituents may result from a variety of pathogenetic factors, such as altered taste acuity, reduced food intake, malabsorption, and the metabolic consequences of concurrent disease. To date, there are only a few studies that have looked concurrently at both nutrition and immunity in the elderly (Chandra, 1985).

The age-related declines in cell-mediated immunity, T-cell number, and thymic factor activity resemble to some extent those seen in protein-calorie malnutrition. Since the elderly are known to be among the most poorly nourished in industrialized countries and since changes in food intake, body composition and protein metabolism are known to occur with advancing age, the possibility arises that altered immune status in the elderly could be ascribed in part to nutritional deficiency. The crucial test for this is to evaluate nutritional and immunologic status before and after deliberate supplementation. Among a group of 51 subjects above the age of 60 years, we found evidence of nutritional deficiencies in 21 and studied the nutritional and immunologic status of these malnourished elderly before and after 8 weeks of dietary supplementation (Chandra et al. 1982). There was clinical, anthropometric, biochemical, and haematologic evidence of malnutrition. Serum ferritin was low in 4 and plasma zinc low in 7. After 8 weeks of supplementation, there was improvement in delayed hypersensitivity response, number of T-cells, and lymphocyte proliferation response to mitogens and antigens (Chandra et al. 1982). Individual nutrient deficiencies are not uncommon in the elderly. Thus correction of these may also be expected to improve immune responses.

To date, the causal contribution of altered nutritional status in the elderly to deficits in their immune responses has not been adequately evaluated and assessed quantitatively (Chandra, 1985, 1989). Our data indicate that a causal relationship does exist between undernutrition and impaired immunity in many elderly individuals and that this is a correctable abnormality in the majority. It is not clear whether maintenance of good nutrition and improved immuncompetence will alter morbidity and longevity. Obviously, long-term prospective studies are required to answer these questions.

FINAL REMARKS

Investigations on the topic of nutrition-immunity interactions have yielded many interesting observations that have both fundamental significance and practical applications. A few examples are given here. The outcome of surgical patients can be predicted on the basis of preoperative assessment of nutritional status and of immunocompetence (Table 1). Moreover, tests of immunocompetence can be used as sensitive functional indices of nutritional status (Table 2). Further, response to prophylactic immunization can be improved if nutritional support is provided before, and even after, the administration of the vaccine. The incidence of opportunistic infections can be reduced when nutritional care is provided at the same time as medical treatment. However, one should caution against the use of megadoses of nutrients since their excessive intake can also impair immune responses.

TABLE 1. Reliability of preoperative anergy (absence of delayed cutaneous hypersensitivity) to predict postoperative mortality and associated complications.

Outcome	Sensitivity	Specificity	Positive predictive value	Negative predictive value
Sepsis	61 (50-80)	74 (62-90)	28 (14-39)	92 (84-97)
Mortality	49 (37-60)	74 (60-88)	25 (21-29)	91 (90-92)

Data are shown as percentage mean (and range) of values obtained from several reported studies, largely based on adults undergoing major elective surgery for a variety of primary diagnoses.

TABLE 2. Methods of nutritional assessment.

Clinical	Biochemical
Dietary intake	Serum albumin
Physical examination	Serum transferrin
	Creatinine/height ratio
	Zinc
Anthropometric	**Immunologic**
Weight-for-height	Delayed hypersensitivity
Mid-upper arm circumference	Lymphocyte count
Skin-fold thickness	Terminaltransferase activity
	T lymphocyte number
Hematologic	**Miscellaneous**
Hemoglobin	Hand grip strength
Red cell morphology	Dark adaptation
Ferritin	Taste acuity

REFERENCES

Beisel, W. R., 1982, Single nutrients and immunity, **Am. J. Clin Nutr.**, 35:417.

Bendich, A., 1988, Antioxidant vitamins and immune responses, *in*: **"Nutrition and Immunology,"** R. K. Chandra, ed., Alan R. Liss, New York.

Chandra, R. K., 1972, Immunocompetence in undernutrition, **J. Pediatr.**, 81:1194.

Chandra, R. K., 1983, Nutrition, immunity and infection. Present knowledge and future directions, **Lancet i,** 688.

Chandra, R. K., 1985, Antibody response in first and second generation offspring of nutritionally deprived rats, **Science**, 190:189.

Chandra, R. K., ed., 1985, **"Nutrition, Immunity and Illness in the Elderly,"** Pergamon Press, New York.

Chandra, R. K., ed., 1988, **"Nutrition and Immunology,"** Alan R. Liss, New York.

Chandra, R. K., 1989, Nutritional regulation of immunity and risk of infection in old age, **Immunology** (in press).

Chandra, R. K., and Newberne, P. M., 1977, **"Nutrition, Immunity and Infection. Mechanisms of Interactions,"** Plenum Publishing Corporation, New York.

Chandra, R. K., Joshi, P., Au, B., Woodford, G., Chandra, S., 1982, Nutrition and immunocompetence of the elderly. Effect of short-term nutritional supplementation on cell-mediated immunity and lymphocyte subsets, **Nutr. Res.,** 2:223.

Citation Classic, 1987, Immunocompetence in undernutrition, **Current Contents,** 30:15.

Dinarello, C. A., and Mier, J. W., 1987, Lymphokines, **New Engl. J. Med.,** 317:940.

Hoffman-Goetz, L., 1988, Lymphokines and monokines in protein-energy malnutrition, *in*: **"Nutrition and Immunology,"** R. K. Chandra, ed. Alan R Liss, New York.

Keusch, G., Wilson, C. S., Waksal, S. D., 1983, Nutrition, host defenses, and the lymphoid system, **Arch. Host Def. Mech.,** 2:275.

McMurray, D. N., 1984, Cell-mediated immunity in nutritional deficiency, **Prog. Food Nutr. Sci.,** 8:193.

THE ROLE OF ANTIOXIDANTS IN MODULATING NEUTROPHIL FUNCTIONAL RESPONSES

Laurence A. Boxer

Division of Pediatric Hematology/Oncology
CS Mott Childrens Hospital
University of Michigan
P. O. Box 0238
Ann Arbor, MI 48109

OXIDATIVE METABOLISM OF THE NEUTROPHIL

Prior to phagocytic stimulation the neutrophil consumes only small amounts of oxygen and relies primarily on anaerobic glycolysis for energy, a finding consistent with the paucity of mitochondria in the cells (Cheson et al., 1977). Within seconds after stimulation, however, the neutrophil activates a specialized pathway that results in a 100-fold increase in oxygen consumption (Root and Metcalf, 1977). This pathway culminates in the reaction in which molecular oxygen is reduced to superoxide (O_2-). The superoxide in turn undergo further reactions to form toxic derivatives such as hydrogen peroxide (H_2O_2), hydroxyl radical ($\cdot OH$), organic oxygen radicals, and hypochlorite ion (HOCl), which are then used by the cell to destroy ingested microorganisms (Babior and Crowley, 1983; Tauber et al., 1983). Massive increases in oxygen consumption and oxygen radical production are referred to as the respiratory burst.

The central enzyme in the respiratory burst is a pyridine nucleotide oxidase, NADPH oxidase which resides in the plasma membrane of neutrophils. This enzyme catalyzes the 1-electron reduction of molecular oxygen to the superoxide radical utilizing electrons from NADPH according to the following equation:

$$NADPH + 2O_2 \longrightarrow NADP + 2O_2\text{-} + H+$$

While either NADPH or NADH can serve as a substrate, the former is believed to be the physiologic electron donor.

NADPH is generated by the action of two enzymes in the hexose monophosphate shunt; glucose-6-phosphate dehydrogenase (G6PD) and 6-phosphogluconate dehydrogenase (6PGD). The accumulation of NADP+ following the onset of the burst increases the flow of substrate through the hexose monophosphate shunt by nearly ten-fold regenerating NADPH.

The superoxide generated by the respiratory burst oxidase undergoes a complex series of reactions to yield other products of the respiratory burst. Of these H_2O_2 is also identified in large amounts in suspension of neutrophils that have undergone stimulation. In addition oxidized halogens and oxidizing radicals such as $\cdot OH$ are produced during the phagocytic respiratory burst. With most stimuli, virtually all of the oxygen consumed in the burst is reduced by one electron to O_2- (Babior and Crowley, 1983). The H_2O_2 is then formed by the

Supported in part by NIH AI 20065 HL 31963

dismutation of superoxide utilizing the cytoplasmic enzyme superoxide dismutase according to the following reaction:

$$O_2- + O_2- + 2H^+ ---> H_2O_2 + O_2$$

The hydrogen peroxide generated by the reaction in turn can then be destroyed by either catalase or the glutathione peroxidase-glutathione reductase system (McAllister et al., 1980).

The purpose of the respiratory burst is to generate reactive oxygen compounds which can be employed to kill microorganisms. Although O_2- and H_2O_2 are the two major products of the respiratory burst they are relatively nontoxic except at high concentration. The neutrophil utilizes two pathways for converting these molecules in substantially more toxic species which are then used in microbicidal reactions. One pathway employs H_2O_2 and myeloperoxidase, a 151 kd heme enzyme that is found in azurophil granules in the neutrophils to form hypochlorite ion (OCL-) (Matheson et al., 1981). The OCL- is quite reactive and is capable of oxidizing a wide variety of biological molecules including cytochromes, nucleotides, and sulfhydryl groups (Thomas, 1979). The myeloperoxidase system is effective against a wide variety of bacteria, fungi, mycoplasmas, and viruses. its cytotoxic activity is believed to be mediated by at least two different mechanisms. First, the hypochlorite ion itself directly oxidizes a wide variety of biomolecules including proteins, cytochromes, nucleotides, and sulfhydryl groups. Additionally, hypochlorite ion oxidizes N-terminal amino acids of cellular proteins (McGuire et al., 1982). It is now well established that hypochlorite can react rapidly with α-1-antitrypsin and affect its inactivation (Stolc, 1979). Since α-1-antitrypsin is one of the major regulators of the neutral proteases, its inactivation results in greatly enhanced proteolytic activity in the vicinity of neutrophils. While this reaction may be of limited microbicidal benefit, it may be a factor in mediating tissue damage during inflammation. Hypochlorite ion can also mediate oxidation of an amine to a chloramine. For example, the N-terminal amino acid of a protein can react with OCL- to generate a dichloramine, which then undergoes spontaneous cleavage (Thomas, 1979):

$$NH_2 - CHR - CO - NHR' + 2OCl^- ---> NCl_2 - CHR - CO - NHR' + H_2O$$
$$NCl_2 - CHR - CO- NHR' ---> CO_2 + 2HCl + RC \equiv N + NH_2R'$$

Similar reactions with lipophilic amines such as ammonia or putrescine yield highly toxic compounds which also play a role in neutrophil host defenses. Fortunately, the neutrophil itself is partially protected against damage from OCL- by taurine, a hydrophilic amine which is present in the cytosol and can undergo reaction with OCL- to form a stable nonreactive N-chloramine (Weiss et al., 1982).

In other reactions OH-, a highly reactive species, can be formed by a direct reaction between O_2- and H_2O_2. This reaction known as the Haber-Weiss reaction has been found to be too slow to result in appreciable rates of OH- production in neutrophils. However in the presence of an iron catalyst, the reaction proceeds quickly according to the following schema (Hallwell and Gutteridge, 1986):

$$Fe^{2+} H_2O_2 ---> \quad Fe^{3+} + OH + OH-$$

$$Fe^{3+} O_2- ---> \quad Fe^{2+} + O_2$$

$$NET \quad H_2O_2 + O_2 ---> \quad O_2 + \cdot OH + OH-$$

Hydroxyl ion radical attack usually sets off a radical chain reaction (Hallwell and Gutteridge, 1986). For example, $\cdot$OH can attack membrane lipids by hydrogen atom abstraction leaving behind a carbon radical (lipid$\cdot$) that swiftly combines with oxygen:

$$lipid-H + \cdot OH ---> H_2O + lipid \cdot \tag{1}$$

$$lipid \cdot + O_2 ---> lipid-O_2- \tag{2}$$

Among other reactions, the peroxy radical can attack adjacent lipids in the membrane to propagate the chain reaction:

$$\text{lipid - O}_2\cdot + \text{lipid - H} \tag{3}$$

$$\text{lipid}\cdot + \text{lipid-O}_2\text{H}$$

(lipid peroxide)

$$\text{lipid}\cdot + \text{O}_2 \longrightarrow \text{lipid-O}_2\cdot \tag{4}$$

Additionally, lipid peroxides can fragment to give a wide range of products, including more radicals, hydrocarbon gases and aldehydes that are highly cytotoxic even in small amounts. Recent studies have shown that exposure of hemoglobin to lipid peroxides or to H_2O_2 can cause a release of iron. It is this released iron which can be bound by chelating agents such as desferrioxamine that seems responsible for hemoglobin stimulated radical reactions. For instance hemoglobin found within an inflamed rheumatoid joint could be a source of liberated iron as it is exposed to H_2O_2 generated by activated neutrophils. The released iron could then amplify lipid peroxidation and ·OH radical formation. Another source of iron is lactoferrin. However, iron is not mobilized from lactoferrin until an acidic pH occurs which does not take place on the surface of the neutrophil but potentially lactoferrin or transferrin accumulation in inflamed rheumatoid joint at low pH might be involved in accelerating ·OH radical formation.

Thus the oxygen radicals generated during the respiratory burst are capable of inflicting damage directly on microorganisms and tissues. The importance of oxidizing radicals in the microbicidal system of the neutrophil is underscored by experiments in which they are intercepted before they interact with their targets. Microbicidal activity is inhibited by catalase, by superoxide dismutase or by oxygen radical scavengers such as mannitol. Since the oxygen radicals produced by the neutrophil are capable of damaging not only bacteria but also normal tissues in sites of inflammation, several radical scavengers are being studied as anti-inflammatory agents. Drugs such as desferrioxamine, 2,3-dihydroxybenzoic acid, and α-tocopherol have proven to be highly efficacious in preventing extracellular damage mediated by hydroxyl radicals in model systems (Boxer, 1986). As indicated below there is growing evidence that neutrophils can be activated in vivo by complement or immune complexes in the Adult Respiratory Distress Syndrome (ARDS) and other diseases. In animal models of ARDS, lung damage can be minimized when oxygen radicals scavengers are administered during the course of the disease. Perhaps at some time in the future antioxidants may play a role in ameliorating clinical disorders secondary to extracellular release of neutrophil generated oxidants.

As discussed, above the enzyme responsible for the respiratory burst is a pyridine nucleotide oxidase that catalyzes the one electron reduction of oxygen to O_2-. There is growing evidence that the NADPH oxidase may be composed of several subunits of which one component of the oxidase traverses the membrane completely conducting electrons from the pyridine nucleotides on the cytoplasmic side to oxygen in the external environment (Babior and Crowley, 1983). Extensive studies have been carried out on the stoichiometry and kinetics of the reaction catalyzed by the respiratory burst oxidase. The stoichiometry expected for the reaction is:

$$2O_2 + \text{NADPH} \longrightarrow 2O_2\text{-} + \text{NADPH}^+ + H^+$$

Lipid of some sort appears to be absolutely necessary for oxidase activity. In addition, the oxidase may be composed of at least two subunits (Curnett and Babior, 1987). One well established subunit is a flavoprotein that contains flavin adenine dinucleotide (FAD) as the proshetic group (Curnett and Babior, 1987). Another potential subunit is a b-cytochrome, which is somewhat similar to previously described b-cytochromes such as the one present in mitochrondia. In the neutrophil b-cytochrome has a dual location; most of it is found in the specific granules, but a small amount can be found in the plasma membrane. Upon activation of the cell much of the b-cytochrome in specific granules is transferred to the plasma membrane as a consequence of degranulation that generally accompanies the activation process. Although b-cytochrome translocation occurs readily it is not necessary for activation of the respiratory burst, to occur. On the other hand the endogenous membrane bound b-cytochrome may play a key role in mediating the respiratory burst. The properties of the b-cytochrome have been extensively studied and it appears the b-cytochrome can be readily reduced by O_2-. Its role in oxidase activation is best supported by the cloning of a putative chronic granulomatous disease (CGD) gene on the basis of DNA linkage (Boxer and Morganroth, 1987). As indicated below CGD cells fail to undergo the respiratory burst. From cDNA sequences the X-CGD mRNA has been found to encode a basic polypeptide containing potential N-glycosylated sites and bearing no homology to known proteins. Antibodies have been raised to a synthetic peptide derived from the cDNA sequence. Further studies revealed that the antibody detected a neutrophil protein of about 90 kd which is absent in X-linked CGD. The antisera also reacted with a recently purified b-cytochrome and with a deglycosylated 90 kd material molecular weight of about 50kd. These findings suggest that the X-linked CGD protein is a heavily glycosylated membrane protein that corresponds to a subunit of purified b-cytochrome. Since the X-linked CGD protein does not resemble a known cytochrome by its sequence, these data suggest that one critical role of this protein might be to anchor b-cytochrome into the membrane. Since b-cytochrome is absent in X-linked CGD patients, it is probable that the protein is linked to the primary defect in the failure of the cells to undergo the respiratory burst in CGD.

One of the major thrusts of current phagocyte research is directed toward understanding the biochemical mechanism by which the respiratory burst is activated. Several important advances have been made in our understanding of the process in the past several years. As discussed above, the ultimate step of the activation process is the conversion of NADPH oxidase from an inactive to an active form. There is growing evidence that this activation process is in fact a multistep pathway linking receptor-ligand interactions on the cell membrane with ultimate activation of the NADPH oxidase (Curnette and Babior, 1987). The 15-60 second lag between exposure of the phagocyte to a stimulus and the onset of the respiratory burst is believed attributable to this activation pathway. Since the neutrophil is capable of activating the respiratory burst in response to a remarkably wide variety of stimuli (e.g., opsonized bacteria, chemotactic peptides, immune complexes, unsaturated fatty acid, phorbol esters, fluoride ion, etc.), there are presumably several activation pathway branches which eventually converge and result in oxidase activation.

The identity of the biochemical intermediates and the activation pathway are unknown although a number of attractive candidates have been proposed. These include transient rises in cyclic AMP concentration, calcium mobilization from membrane stores, phospholipid turnover, protein phosphorylation and changes in membrane potential. All of these biochemical changes have the important property that they occur within seconds after stimulation and precede the onset of superoxide production. In addition, all of these biochemical changes are potentially rapidly reversible. Because of the complexity of the neutrophil response to stimuli, however, it has not been possible to link definitively any of these changes with activation of the oxidase. The ultimate resolution of the activation mechanism problem likely depends both on the purification of NADPH oxidase and the establishment of techniques which will enable investigators to manipulate substantially the intracellular milieu of the neutrophil. To achieve that goal, a cell-free system has been developed in which the dormant NADPH oxidase enzyme can be activated under highly controlled conditions (Curnette and Babior, 1987). It has been shown employing arachidonic acid as a stimulus that the dormant NADPH oxidase in unstimulated neutrophils again appears to be entirely confined to the plasma membrane fraction. The addition of a cytosolic factor and arachidonate to the plasma membrane fraction elicits oxidase activity. The identity of the cytosolic factor is unknown but the factor does not appear to be protein kinase C. In other studies, partially purified protein kinase C was shown to activate the NADPH oxidase in a cell free system if ATP, calcium, and phorbol myristate acetate were present. These studies indicate that at least two pathways leading to activation of the NADPH oxidase are possible. Phorbol ester-mediated activation of protein kinase C is required presumably to phosphorylate some regulatory catalytic component of the oxidase complex. For arachidonic-mediated activation a different cytosolic protein is required to facilitate the activation of the oxidase in a cell-free system. The cytosolic factor has been found to be deficient in some patients with autosomal recessive CGD but normal in a series of patients with X-linked CGD (Curnette and Babior, 1987). It is possible then that the autosomal recessive form of CGD may be caused by defective or absent cytosolic factor. A model for activation of the NADPH oxidase linking two different stimuli phorbol myristate acetate and chemotactic peptide to activation of the NADPH oxidase is suggested. In a now familiar theme, one pathway could be activated by the binding of N-formyl chemotactic peptides to their receptor on the neutrophil membrane leading a G protein activator of phospholipase C and generation of the calcium releaser IP_3. The other pathway could arise from generation of diacylgcerol leading to activation of protein kinase C. Potentially the activation of calcium/calmodulin complex by the rise in free calcium could phosphorylate the cytosolic factor whereas protein kinase C could phosphorylate the oxidase itself bypassing the need for a soluble cofactor. The precise change in NADPH oxidase that leads to catalytic activity remains unknown but one possibility might involve assembly of oxidase subunits such as b-cytochrome with flavoprotein following modification of the subunit by phosphorylation. Finally the use of the cell-free system also lends itself to examining once again the role of antioxidants in scavenging respiratory burst products.

The importance of oxidizing radicals to the microbiocidal system of the neutrophil is underscored by experiments in which they are intercepted before they can interact with their targets. Microbiocidal activity is inhibited by catalase, by superoxide dismutase, or by oxygen radical scavengers such as mannitol. Since the oxygen radicals produced by the neutrophils are capable of damaging not only bacteria but also normal tissues in sites of inflammation the oxygen scavengers along α-tocopherol are being studied as anti-inflammatory agents. In animal models of neutrophil-mediated lung damage, pulmonary disease can be minimized when oxygen radical scavengers are administered during the course of the experiment (Boxer and Morganroth, 1987). It would appear then that antioxidants can be employed to manipulate neutrophil function through their ability to scavenge released oxygen by-products as well as attenuate tissue injury inflicted by these products.

Consequence of α-Tocopherol Deficiency on Neutrophils Function

In neonates, both random motility of neutrophils and chemotaxis are diminished (Miller, 1971). Other abnormalities of neutrophil phagocytes have been noted in human newborns. These include the decreased intracellular killing in the presence of a stress or a secondary illness. The etiology of these abnormalities of phagocyte function remain unknown but possibly could relate to differences in membrane properties between newborn and adult neutrophils (Wright et al., 1975). Others have found by treating neutrophils with 2,3-dyhydroxybenzoic acid (2,3-DHB) that chemotactic activity could be improved (Shigeoka et al., 1981). Potentially, the active metabolic state of newborn neutrophils in concern with the relative α-tocopherol deficiency at birth could contribute to alteration in cellular function through membrane damage by predisposing the neutrophil to autooxidant-induced injury (Hill, 1987; Baehner et al., 1977; Oski, 1977). We previously assessed whether α-tocopherol deficiency could affect neutrophil function (Harris et al., 1980). Thus, we rendered weanling rats α-tocopherol deficient and evaluated neutrophil behavior. Following 8 weeks of an α-tocopherol deficient diet, serum α-tocopherol levels were 0.06 ± 0.01 mg/100 ml compared to levels in control rats of 1.17 ± 0.06 mg/100 ml. Neutrophils obtained from the peritoneal cavity of rats revealed that directed cell movement but not random motility was depressed in the α-tocopherol-deficient animals. Similarly, the ingestion of both C3b opsonized or IgG opsonized paraffin oil droplets was significantly compromised by neutrophils obtained from the α-tocopherol-depleted animals (Table 1). We further found an increase by 1.5-fold of the amount of peroxidized lipid membrane as assayed by the formation of malonyldialdehyde in the α-tocopherol-depleted neutrophil compared to control. Further studies revealed that α-tocopherol-deficient neutrophils were metabolically more active than controls as evidenced by an increase in oxygen consumption and release of hydrogen peroxide into the extracellular medium. We interpreted these findings based on the hypothesis that the membrane-associated NADPH oxidase responsible for the consumption of oxygen and generation of oxygen byproducts in the α-tocopherol deficient neutrophil was situated in an abnormal lipid milieu which in turn facilitated greater activation of the enzyme and generation of auto-toxic concentrations of hydrogen peroxide.

To appreciate the consequence of α-tocopherol depletion, it is necessary to outline the role of the vitamin in inhibiting the oxidation of unsaturated fatty acids. The oxidation of fatty acids is dependent on presence of double bonds which are present in the unsaturated fatty acids but not in the saturated fats (Pryor, 1982). Such auto-oxidation of an unsaturated fatty acid like linoleic acid has been postulated to occur as follows: 1) loss of a hydrogen ion atom from the carbon in the alpha position of the double bond; 2) nonenzymatic oxidation of the carbon atom to a free peroxidic radical; and 3) reaction of the unsaturable lipid-free radicals with other unsaturated fatty acid molecules which generate lipid hydroperoxides, and more lipid-free radicals. Because of its capacity to scavenge lipid-free radicals through the phenolic group found on the chroman ring of the molecule (Figure 1), α-tocopherol prevents the oxidation reactions described above. In turn the fatty acid and side chain of the α-tocopherol molecule accounts for the solubility in lipids thereby rendering the vitamin useful in preventing oxidation within the plasma membrane.

Table 1. C3b and Fc Receptor - Mediated Ingestion of Opsonized Paraffin Oil Droplets by α-Tocopherol Deficient Rat PMN

	Control	α-Tocopherol
Deficient		
C3b receptor-mediated ingestion*	0.093 ± 0.09	0.051 ± 0.071
Fc reception-mediated ingestion*	0.0976 ± 0.081	0.052 ± 0.029

mg* paraffin oil/10^7 neutrophils/min

+Each sample represents the mean ± S.D. of at least eight samples. All values of controls compared to deficient cells are significant to a p values of < .0005.

Function of Normal Neutrophils in the Presence of Antioxidants and of CGD Neutrophils

Clearly, α-tocopherol deficiency lead to sufficient damage which compromises neutrophil function. On the other hand, what are the consequences of supplementing the diet with α-tocopherol on neutrophil function? To answer the question, we administered 1600 units of α tocopherol to adult volunteers for a fortnight (Baehner et al., 1977). Serum α-tocopherol levels rose from a range of 1.12 ± 0.16 to 1.82 ± 0.04 mg/100 ml. As noted in table 2, the rate of uptake of C3-opsonized lipopolysaccharide-coated paraffin oil droplets increased from 0.060 ± 0.015 mg/10^7 neutrophils/min before α-tocopherol supplementation to values of 0.076 ± 0.009 mg/10^7 neutrophils/min following treatment. On the other hand, despite the greater ability to phagocytose particles the α-tocopherol-replete neutrophils showed a mildly reduced ability to kill bacteria (Baehner et al., 1977). The extent of this acquired abnormality was far less than that observed for neutrophils in patients with chronic granulomatous disease. These results suggest that the decreased release of hydrogen peroxide by α-tocopherol-replete neutrophils may result in less oxidative damage to their membranes so that they are more efficient in ingestion of C3-opsonized lipopolysaccharide-coated paraffin oil droplets but that the failure to produce this reactive species also serves to protect ingested microorganisms from the peroxide-mediated attack which results in their demise.

We found that α-tocopherol-replete neutrophil released superoxide normally to the extracellular medium but the neutrophils released only 45% of the amount of hydrogen peroxide as control. Alpha-tocopherol also served to scavenge endogenously generated hydrogen peroxide within the cell. To assess the availability of hydrogen peroxide within the cell, ^{14}1-C glucose oxidation through the hexose monophosphate shunt, which depends upon generation of intracellular hydrogen peroxide, was determined. In the control neutrophils, hydrogen peroxide dependent 1-^{14}C-glucose oxidation increased by 4.4-fold during phagocytosis of latex particles over 15 minutes compared to a 2.9-fold increase in the α-tocopherol-replete neutrophil.

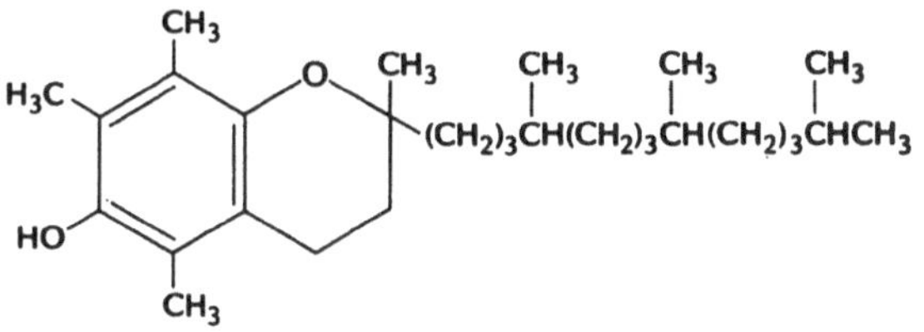

Figure 1. Structure of α-tocopherol. The phenotic group on the chronan ring (left) permits reactivity with free radicals, whereas the fatty acid chain allows for lipid solubility.

Table 2. Rate of Uptake of C_3-coated Lipopolysaccharide Paraffin Oil Droplets[*]

Treatment	Ingestion Rate (mg paraffin oil/10^7 neutrophils/min)	P Value
Control	0.060 ± 0.015	<.025
α-Tocopherol replete+	0.076 ± 0.009	NS
10^{-3}M Benzoic Acid	0.066 ± 0.018	<.002
10^{-3}M 2,3-DHB	0.092 ± 0.014	<.002
10^{-4}M 2,3-DHB	0.087 ± 0.050	<.025
10^{-5}M 2,3-DHB	0.070 ± 0.011	NS

*Results are expressed as mean $\pm$ SEM. The P values are based on student's T test with paired values and are compared to control.

+Volunteers were studied before and during ingestion of α-tocopherol 1,600 IU/day for 14 days; the peripheral blood neutrophils were then examined for their phagocytic ability. The remainder of the drug studies with benzoic and 2,3-dihydroxygenzoic acid (2,3-DHB) were conducted by adding the drug to normal neutrophils *in vitro*

Although hydrogen peroxide released both extracellularly and intracellularly was reduced in the α-tocopherol-replete neutrophil, the cells were able to utilize oxygen normally.

The site of hydrogen peroxide produced by NADPH oxidase was determined in neutrophils obtained from the blood of volunteers receiving 1600 units of α-tocopherol daily (Figure 2). For these studies, neutrophils were incubated in the presence of NADPH and cerious chloride and allowed to generate hydrogen peroxide by phagocytosing opsonized zymosan particles (Butterick et al., 1983). The hydrogen peroxide formed from NADPH reacts with cerium to form an electron-dense reaction product cerium perhydroxide, detected either visually or by x-ray microanalysis. The equation for the reaction is as follows:

1. $NADPH + O_2 + H^+$ oxidase---> $NADPH + H_2O_2$

2. $2H_2O_2 + CeCL_3$ ---> $Ce(OH)_2OOH + H^+$.

The reaction product was identified on the plasma membrane as well as within phagocytic vesicle membranes surrounding ingested zymosan when NADPH was employed as the substrate with normal neutrophils. When neutrophils obtained from the volunteers receiving α-tocopherol were incubated with NADPH, no reaction product developed either within the phagocytic vesicles or on the plasma membrane suggesting that α-tocopherol is capable of scavenging hydrogen peroxide needed for the formation of this reaction product.

25

Because we observed that α-tocopherol-replete cells destroyed bacteria less efficiently than control, we turned our attention to another scavenger of hydrogen peroxide, 2,3DHB in an effort to better manipulate the autooxidative reactions engendered by activated neutrophils (Graziano et al., 1974). This drug has been identified as an orally effective iron chelator (Graziano et al., 1976). It is an aromatic compound having para-situated hydroxy groups which have the capability of reacting with free radicals to form longer-lived semiquinones. The semiquinones subsequently react with another free radical to form quinones. A scheme whereby 2,3-DHB might react is depicted (Figure 3).

We found the neutrophils phagocytosed at a faster rate and augmented their directed cell movement in the presence of 2,3-DHB *in vitro* (Table 2). The concentrations of 2,3-DHB, we employed which were needed to alter neutrophil functions *in vitro* were readily attainable *in vivo* (Boxer et al., 1978). The administration of 25 mg/kg of 2,3-DHB for use in iron-chelation chemotherapy has resulted in a plasma concentration 0.5 - 0.6 mM. Unlike α-tocopherol, removal of 2,3-DHB from the extracellular media abrogated the effect of the drug. Alpha-tocopherol is incorporated into the membrane lipid bilayer and can reduce intracellular levels of hydrogen peroxide sufficiently to impair bacterial killing; whereas 2,3-DHB failed to alter the bactericidal capacity of the neutrophil. Hence, 2,3-DHB may be a more useful drug in some situations in enhancing neutrophil motile responses without compromising bactericidal potency.

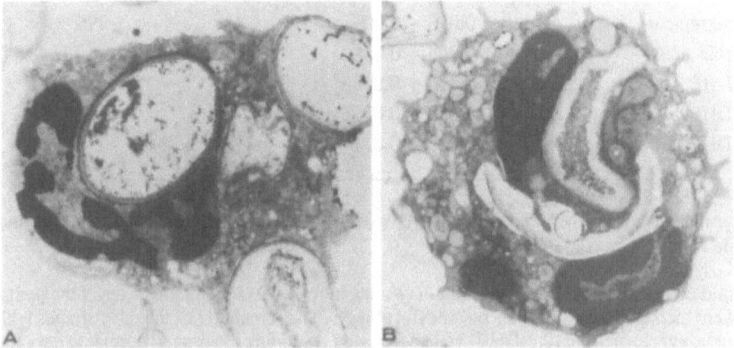

Figure 2. Ultrastructural localization of H_2O_2 in human neutrophils from volunteers receiving 1,600 units vitamin E daily after phagocytosis of opsonized zymosan. Panel A indicates results with 0.71mM NADPH. Note the presence of reaction product on the membrane lining the phagocytes vesicles and on the plasma membrane. Panel B indicates results obtained with 0.71mM NADPH using neutrophils obtained from volunteers receiving vitamin E. Note that there is marked attenuation of the reaction product. Magnification X8000 Figure taken from reference 24.

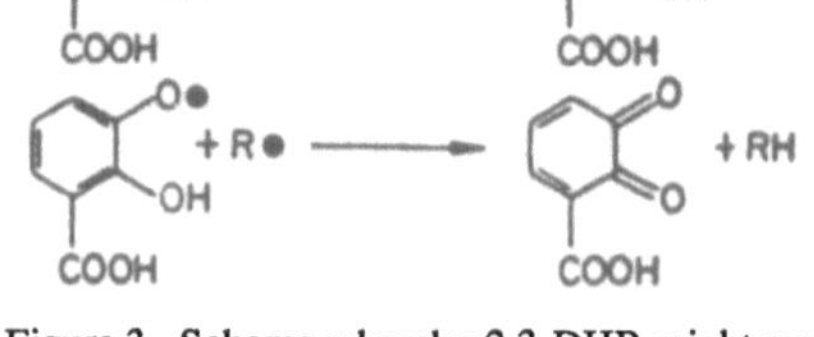

Figure 3. Scheme whereby 2,3-DHB might react.

To better assess the role of antioxidants in modulating neutrophil phagocytosis, we have developed a novel optical microscopic methodology to follow neutrophil-mediated effector function at the level of individual cells. We found that the oxidation of intracellular hemoglobin could be observed by bright-field microscopy using illumination at 430nM (Francis et al., 1978). Exposure of hemoglobin to superoxide anion decreased the intensity of the Soret band

and shifted it to lower wave-lengths. We were thus able to follow the oxidation of the erythrocyte hemoglobin by employing a 430 nM band pass interference filter in conjunction with bright-field optical microscopy. We found that native hemoglobin absorbed intensely and appeared dark while oxidized hemoglobin became transparent as we followed the sequential oxidation of bound targets by neutrophils (Figure 4). Cytosolic or membrane compartments of sheep erythrocytes were also labeled with eosin Y or fluorescein isothiocynate, respectively.

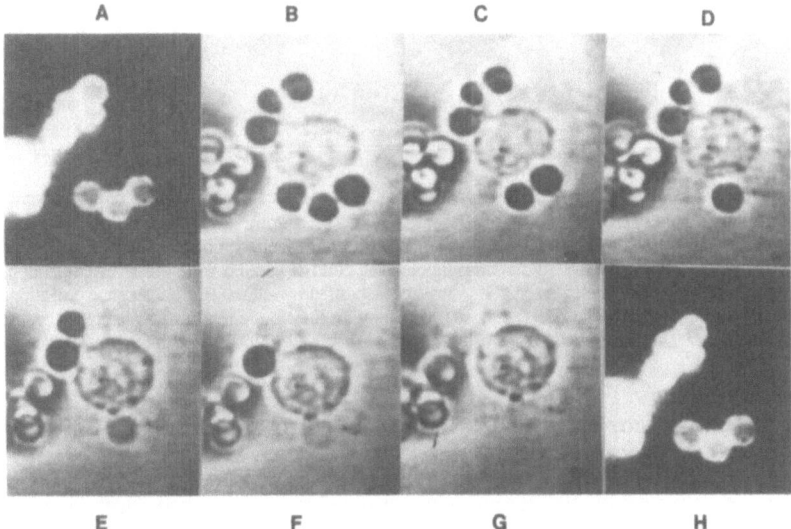

Figure 4. Temporal analysis of erythrocyte oxidation by neutrophils. Roughly 0.5 x 10^6 neutrophils were allowed to adhere to glass coverslips at 37°C. IgG fluorescein labeled red cells (0.5 x 10^7) were added to adherent cells. The coverslips were secured to prewarmed microscope stage at a nominal temperature of 37°C. Cell images were recorded on a time-lapse videorecorder. Bright-field microscopy experiments were conducted using a 430 nm interference filter to provide excitation at the soret band. Attached erythrocyte targets are found to sequentially lose their contrast at 430 nm (b-g) but remain intact as shown by fluorescence microscopy (x 1,200) (a,h). The time from panel b to h is 1:28. Individual frames are separated by 6 to 27 sec. Taken from reference 28.

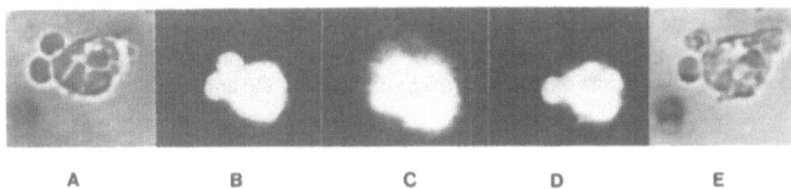

Figure 5. Extracellular lysis of fluorescent erythrocytes. IgG-opsonized eosin-labeled red cells were added to adherent neutrophils as described in the legend of Figure 4. Cells were maintained at about 37°C on a warmed microscopy stage. Five internalized and two bound eosin-labeled red cells can be seen in panel a. The bound cell indicated by the arrow in panel b was lysed in panel c. Lysis was observed as a burst of fluorescence emission around the periphery of the target lysed. The cells were devoid of fluorescence (d) and cell contrast at 430 nm was diminished (e). The time period from a to e was 39 sec. (x 1,200). Taken from reference 28.

27

Time-dependent time studies of erythrolysis show that antibody coated red cell targets were lysed extra-and intracellularly following incubation with neutrophils (Figures 5,6). This provided a simple and efficient microscopical method to measure target cell rupture. Using chronic granulomatous disease (GCD) neutrophils which do not generate superoxide we found that the CGD neutrophils were capable of ingesting targets but failed to lyse erythrocytes by oxidative mechanisms (Figure 7). These sets of experiments now provide a fresh approach to the study of phagocyte effector function including a novel method to assess the role of generated oxidants intra- and extracellularly.

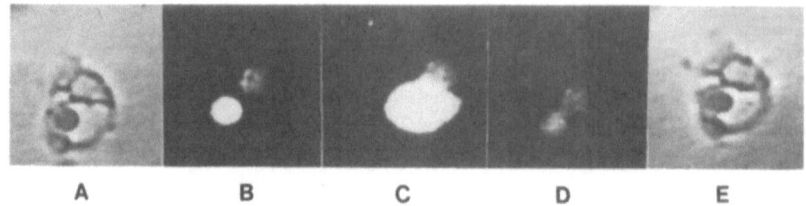

A   B   C   D   E

Figure 6. Intracellular lysis of fluorescent erythrocyte targets. IgG-opsonized eosin-labeled red cells were ingested by adherent neutrophils after 10 min. at 37°C (see Figure 4). Cells were observed on a warmed microscopy stage. Two large phagosomal compartments can be seen using bright-field microscopy (a). Lysis of the bright eosin-labeled red cells of panel b were observed in panel c. Lysis of an internalized eosin-labeled red cells were accompanied by a burst of fluorescence emission. The burst of fluorescence was restricted to one phagosome; it does not enter the adjacent phagosome compartment. The time period from panel a to e is 34 sec. (x 1,200). Taken from reference 28.

Effect on Alpha-Tocopherol on Neutrophil Function in a Patient with Glutathione Synthetase Deficiency

A number of patients with glutathione synthetase deficiency have been described in which reduced glutathione in the red cells was markedly depressed and associated with chronic nonspherocytic hemolytic anemia and hemolytic crises following drug ingestion (Boivin et al., 1966; Mohler et al., 1970). The inheritance follows an autosomal recessive pattern. In the oxoprolinuric variant of glutathione synthetase deficiency, nucleated cells such as leukocytes and fibroblasts have been found to have depressed glutathione levels (Spielberg et al., 1978). In one child suffering from neutropenia following episodes of recurrent otitis media neutrophil levels of glutathione synthetase were 5% of normal and levels of glutathione at 10-20% of normal (Spielberg et al., 1979). While the chemotaxis of phagocytes was normal, killing of ingested staphylococci was moderately impaired. Hydrogen peroxide production by phagocytosing neutrophils was 150% greater than control while iodination and microtubule assembly were markedly impaired. After three months of oral α-tocopherol (30 U/kg/day), the phagocytosing neutrophils released normal amount of hydrogen peroxide and iodinated protein assembled microtubules and killed *Staphylococcus aureus* normally (Boxer et al., 1979). Cellular levels of glutathione, however, did not rise. These findings suggest that the patient's neutrophils could not appropriately detoxify hydrogen peroxide which then led to microtubule dysfunction and impaired degranulation of lysosomes into phagosomes; thus compromising iodination and bacterial killing. It would appear that α-tocopherol protected the glutathione synthetase-deficient cells against oxidative damage implicating the anti-oxidant behavior of this enzyme.

Attenuation of Lung Damage in Animal Models by Antioxidants

The pathophysiologic mechanisms underlying ARDS are an area of intense research interest (Boxer and Morganroth, 1987). Indirect evidence exists regarding the ability of the neutrophil to produce acute lung injury by elaboration of toxic oxygen metabolites (Till et al., 1982; Schrachraufstatter et al., 1984). Despite our accumulating knowledge of the mechanism of ARDS, little has changed regarding pharmacologic management of the disease. While compounds such as corticosteroids and inhibitors of arachidonic acid metabolism have shown promise in animal studies, confirmation of these findings in human diseases prove difficult and controversial (Brigham et al., 1981; Perkowski et al., 1983). In an effort to discover alternative methods, we chose to evaluate the ability of 2,3-DHB to prevent acute lung injury (Baldwin et al., 1985).

Various models have been utilized to study acute lung injury. Many have implicated neutrophil-derived oxygen metabolites as a cause of the damage (Till et al., 1982; Till et al., 1985). One such model involves infusing cobra venom (naja naja) factor into the rat (Till et al., 1982).The lung injury in this animal model appears to be mediated by neutrophil-generated oxygen metabolites since both neutrophil depletion or prior administration of oxygen metabolite scavengers have a protective effect. Changes observed after treatment of the rat with cobra venom factor (CVF) resembles those seen in adult rats. Those changes include: 1) complement activation with formation of chemotactic fragments; 2) histologic evidence of neutrophil accumulation into lung capillaries and alveolar spaces; 3) high permeability pulmonary edema assessed by the leakage of radiolabeled albumin from the vascular space to the alveolar insterstitium; and 4) histologic evidence of pulmonary edema, alveolar hemorrhage, and fibrin deposition. Using the CVF model of complement and neutrophil-mediated lung injury of the rat, we tested the hypothesis that 2,3-DHB could protect the lung from such injury.

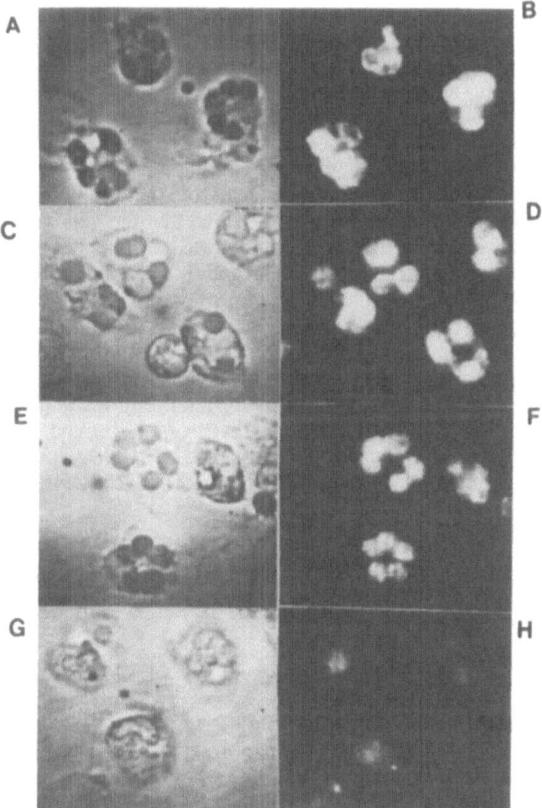

Figure 7. Defective intracellular lysis of IgG-opsonized erythrocyte targets by CGD neutrophils. Neutrophils in suspension at a final concentration of 1×10^6 cells/ml were incubated with eosin-labeled red cells at a target-to-effector ratio of 10:1 for 10 min. at 37°C. Extracellular eosin-labeled red cells were removed by a 10 sec. hemolysis treatment. Phagocytosed eosin-labeled red cells can be seen in control (a,b) and CGD (c,d) neutrophils. After incubation for 10 min. at 37° neutrophils (1×10^6/ml) from control (a,b) and CGD patients (c,d) are observed with intra-cellular eosin-labeled red cells. After 30 min. at 37°C fluorescent erythrocyte targets remain visible in CGD neutrophils (e,f) while normal control neutrophils possess only a dim cell-associated fluorescence (g,h). Panel a,c,e, and g are bright-field while b,d,f, and h are fluorescence photomicrographs. (x 1,200) Taken from reference 28.

Using a permeability index that measures the amount of intravenously administered ^{125}I-albumin that accumulates in lung tissue, we found that pretreatment with 2,3-DHB reduced (p < 0.05) lung injury in CVF-treated rats in a dose-dependent manner (Table 3). Morphometric analysis of lung tissue indicated that the protection by 2,3-DHB was not caused by CVF-induced neutrophil sequestration within lung vasculature. Because iron-saturated 2,3-DHB did not attenuate lung injury, and because *in vitro* experiments demonstrated that 2,3-DHB inhibited iron-hydrogen peroxide-induced peroxidation of phospholipid liposomes, we suggested that 2,3-DHB may be protecting the lung via chelation of iron as well as serving to protect the lung from hydrogen peroxide-mediated damage.

Experimental thermal injury of rat skin (70C: 30 secs) results also in activation of complement which leads to the appearance of C5 related chemotactic activity in the serum, transient neutropenia, and accumulation of neutrophils in pulmonary capillaries (Till et al., 1985). These events ultimately lead to acute lung injury similar to that observed following the infusion of CVF into rats. Protection from acute lung injury following remote thermal injury (involving skin) can be shown by depleting animals of complement or blood neutrophils or by systemic treatment of animals with combinations of catalase and superoxide dismutase (SOD). The protective effects of catalase and SOD provide strong evidence that oxygen-derived free radicals released from complement-activated neutrophils are important mediators of lung injury secondary to skin burns. The nature of the oxygen species appear to involve hydrogen peroxide and perhaps ·OH. In recent studies it was found that acute lung injury could be prevented by treatment of thermally injured rats with the anti-oxidant α-tocopherol or with scavengers of ·OH such as 2,3-DHB (Table 3). The role of iron is not completely understood in the development of tissue injury secondary to skin burns but possibly could be explained by its essential goal in the classic iron-dependent Fenton reaction in which hydroxyl radical is formed from hydrogen peroxide.

Hydroxyl radical is highly reactive and can lead to generation of lipid peroxides. In the studies involving thermal injury, evidence was provided that the generation of lipid peroxidation products which are detectable in burned skin, lung tissue and plasma may be dependent on hydroxyl release from complement-activated neutrophil. The authors found that both neutrophil depletion and pretreatment with ·OH scavengers almost completely prevented the appearance of conjugated dienes in plasma of thermally injured animals. They also found the hydroxyl radical scavengers and α-tocopherol decreased the plasma levels of conjugated dienes suggesting that lipid peroxidation products are in some manner correlated with both the skin injury and lung injury. These findings provide some suggestions for interventions that might be useful in clinical situations where there is reason to believe that oxygen-derived free radicals may be responsible for tissue injury. One such clinical situation is ARDS where increased amounts of oxidants are found in the breath of patients with the disorder (Baldwin et al., 1986).

Table 3. Effect of 2,3-Dihydroxybenzoic acid and Alpha-Tocopherol on Lung Vascular Permeability.

Stimulus	Treatment	Lung Vascular Permeability	
		Index	P Value
Saline	None	0.018 ± 0.02	-
CVF	None	0.59 ± 0.06	-
CVF	2,3-DHB (100 mg/kg)	0.20 ± 0.05	p < .001
Skin Burn	None	0.68 ± 0.06	-
Skin Burn	2,3-DHB (100 mg/kg)	0.32 ± 0.01	p < .001
Skin Burn	α-Tocopherol	0.34 ± 0.04	p < .01

Lung vascular permeability index values were determined at 30 minutes or 3 hours following infusion of cobra venom factor (CVF) or following thermal injury, respectively.

Potential Toxicity of High Supplemental Levels of α-tocopherol

Among the events accompanying phagocytosis by neutrophils are: 1) recognition of the target by cell surface receptors, 2) triggering of the respiratory burst and granule release; and 3) endocytic engulfment of the target (Curnette and Boxer, 1987). Recently the structural entities participating in Fc and C3b receptor-mediated recognition have been identified (O'Shea et al., 1985; Kurlander and Batker, 1982; Fleit et al., 1982). However, the physiologic transduction mechanisms mediating the broad spectrum of functional changes are not known. Trans-membrane ion fluxes may constitute one early step in metabolic signaling during antibody-phagocytosis (Young et al., 1983). Arachidonic acid metabolism may be triggered by a phosphalipase activity found in the Fc receptors (Suzuki et al., 1982). Additionally, it has been suggested that a cell surface sulfhydryl group participating in a sulfhydryl-oxidation reaction may play a large role in IgG-dependent endocytic triggering (Petty, 1973). We have hypothesized that release of oxidative by-products by phagocytes may constitute a driving force in the IgG-mediated endocytosis of targets. To test this hypothesis, we challenged neutrophil obtained from patients with CGD with immune complexes (Petty et al., 1988). We found that the neutrophils obtained from the patients of GCD were unable to ingest immune complexes where they were able to bind and ingest the immune complexes if they were challenged with complexes capable of generating hydrogen peroxide. In contrast, normal neutrophils were able to ingest immune complexes without difficulty, but they had a markedly impaired ability to ingest immune complexes in the presence of catalase, a scavenger of hydrogen peroxide. We suggest then that the deficiency in antibody-dependent endocytosis is secondary to the defective respiratory burst of neutrophils obtained from patients with chronic granulomatous disease.

These observations suggest that prolonged ingestion of high dose α-tocopherol could lead to selected inhibition of the neutrophil's ability to clear immune complexes. Clinically, one study of premature infants treated prophylactically with high dose α-tocopherol for prevention of retinopathy revealed the infants were at risk of developing sepsis and necrotizing enterocolitis (Johnson et al., 1985). These particular clinical observations coupled with the *in vitro* observations relating potential toxicity to high dose α-tocopherol suggest that the vitamin needs to be administered in moderation.

In summary it is clear that neutrophils and their oxidative by-products have the capability to participate in inflammatory reactions. Further research is required to ascertain the potential benefits and risks arising from the use of antioxidants such as α-tocopherol in modulating the neutrophil respiratory burst in the clinical setting.

REFERENCES

Babior, B. M., and Crowley, C. A., 1983, Chronic granulomatous disease and other disorders of oxidative killing by phagocytes, in: "Metabolic Basis of Inherited Disease," J. B. Stanbury, J. B. Wyngaarden, D. S. Frederickson, eds., McGraw-Hill Book Co., New York.

Baehner, R. L., Boxer, L. A., Allen, J. M., and Davis, J., 1977, Autooxidation as a basis for altered function by polymorphonuclear leukocytes, **Blood**, 50:327.

Baldwin, S. R., Simon, R. H., Boxer, L. A., Till, G. O., and Kunkel, R. G., 1985, Attenuation by 2,3-dihydroxybenzoic acid of complement and neutrophil mediated acute lung injury in the rat, **Am. Rev. Resp. Disease**, 132:1288.

Baldwin, S. R., Simon, R. H., Grumm, C. M., Ketai, L. H., Boxer, L. A., and Devall, L. J., 1986, Oxidant activity in the expired breath of patients with the adult respiratory distress syndrome, **Lancet**, 1:11.

Boivin, P., Galand, C., Andre, R., and Debray, J., 1966, Anemes hemolytique congenitales avec deficit esole en glutathion reduit par deficit en glutathione synthetase, **Now. Rev. Fr. Hematol.**, 6:859.

Boxer, L. A., 1986, Regulation of phagocyte function by α-tocopherol, **Proc. Nutr. Soc.** 45:333.

Boxer, L. A., Allen, J. M., and Baehner, R. L., 1978, Potentiation of polymorphonuclear leukocyte motile function by 2,3-dihydroxygenzoic acid, **J. Lab. Clin. Med.**, 92:730.

Boxer, L. A., and Morganroth, M. L., 1987, Neutrophil function disorders, **Disease-A-Month**, 13:683.

Boxer, L. A., Oliver, J. M., Spielberg, S. P., Allen, J. M., and Schulman, J. D., 1979, Protection of granulocytes by vitamin E in glutathione synthetase deficiency, **New Eng. J. Med.**, 301:901.

Brigham, K. L., Bowers, R. E., and McKeen, C. R., 1981, Methylprednisolone prevention of increased lung vascular permeability following endotoxemia in sheep, **J. Clin. Invest.**, 68:13.

Butterick, C. J., Baehner, R. L., Boxer, L. A., and Jersild, R. A. Jr., 1983, Vitamin E: A selective inhibitor of the NADPH oxido-reductase enzyme system in human granulocytes. **Am. J. Pathol.**, 112:287.

Cheson, B. D., Curnette, J. T., and Babior, B. M., 1977, The oxidative killing mechanism of the neutrophil, **Clin. Immunol.**, 3:1.

Curnette, J. T., and Babior, B. M., 1987, Chronic granulomatous disease, *in*: **"Advances in Human Genetics,"** H. Harris and K. Hirschborn, eds., Plenum Publishing Co., New York.

Curnette, J. T., and Boxer, L. A., 1987, Disorders of granulopoiesis and granulocyte function, *in*: **"Hematology of Infancy and Childhood,"** 3rd Edition, D. G. Nathan and F. A. Oski, eds., W. B. Saunders Co., Philadelphia.

Fleit, H. B., Wright, S. D., Unkeless, J. C., 1982, Human neutrophil Fc receptor distribution and structure, **Proc. Natl. Acad. Sci. USA,** 79:3275.

Francis, J. W., Boxer, L. A., and Petty, H. R., 1988, Optical microscopy of antibody dependent phagocytes and lysis of erythrocytes by living normal and chronic granulomatous disease neutrophil: A role of superoxide anions in extra- and intracellular lysis, **J. Cell Physiol.**135:1.

Graziano, J. H., Grady, R. W., and Cerami, A., 1974, The identification of 2,3-dihydroxygenzoic acid as a potentially useful iron-chelating drug, **J. Pharmacol. Exp. Ther.**, 190:570.

Graziano, J. H., Miller, D., Grady, R. W., and Cerami, A., 1976, Inhibition of membrane peroxidation in thalessemic erythrocytes by 2,3-dihydroxybenzoic acid, **Brit. J. Haemotol.**, 32:351.

Hallwell, B., and Gutteridge, J. M. C., 1986, Iron and free radical reactions: Two aspects of antioxidant protection, **Trends in Biochem. Sci.**, 11:372.

Harris, R. E., Boxer, L. A., and Baehner, R. L., 1980, Consequences of vitamin E deficiency on the phagocytic and oxidative function of the rat polymorphonuclear leukocyte, **Blood**, 55:338.

Hill, H. R., 1987, Biochemical, structural, and functional abnormalities of polymorphonuclear leukocytes in the neonate, **Pediatr. Res.**, 22:375.

Johnson, L., Bowen, F. W., Jr., Abbasi, S., Porat, R., Stahl, G., Peckham, G., Delivoria-Popadopoulos, M., Quinn, G., and Shaffer, D., 1985, Relationship of prolonged pharmacologic serum levels of vitamin E to incidence of sepsis and necrotizing enterocolitis in infants with birth weight 1,500 grams or less, **Pediatrics**, 75:619.

Kurlander, R. J., and Batker, J., 1982, The binding of human immunoglobulin GI monomer and small covalently cross-linked polymer of immunoglobulin GI to human peripheral blood monocytes and polymorphonuclear leukocytes, **J. Clin. Invest.**, 69:1.

Matheson, N. R., Wong, P. S., and Travis, J., 1981, Isolation and properties of human neutrophil myeloperoxidase, **Biochemistry**, 20:325.

McAllister, J. A., Harris, R. E., Baehner, R. L., and Boxer, L. A., 1980, Alteration of microtubule function in glutathione peroxidase deficient polymorphonuclear leukocytes, **J. Reticulo. Soc.**, 27:59.

McGuire, W. W., Spragg, R. G., Cohen, A. M., and Cochrane, C., 1982, Studies on the pathogenesis of the adult respiratory distress syndrome, **J. Clin. Invest.**, 69:543.

Miller, M. E., 1971, Chemotactic function in the human neonate: Humoral and cellular aspects, **Pediatr. Res.**, 5:487.

Mohler, D. N., Majerus, P. W., Minnich, V., Hess, C. E., and Garrick, M. D., 1970, Glutathione synthetase deficiency as a cause of hereditary hemolytic disease, **New Eng. J. Med.**, 283:1253.

O'Shea, J. J., Brown, E. J., Seligmann, B. E., Metcalf, J. A., Franks, M. M., and Gallin, J. E., 1985, Evidence for distinct intracellular pools of reception for C3b and C3bi in human neutrophils, **J. Immunol.**, 134:2580.

Oski, F. A., 1977, Metabolism and physiologic roles of vitamin E, **Hosp. Practice**, 12:79.

Perkowski, S. Z., Havill, A. M., Flynn, J. T., Gee, M. H., 1983, Role of intrapulmonary release of eicosanoids and superoxide anion as mediators of pulmonary dysfunction and endothelial injury in sheep with intermittent complement activation, **Circ. Res.**, 574.

Petty, H. R., 1973, Specific inhibition of macrophage antibody dependent endocytosis by p-Chloromercuribenzene sulfonic acid: Identification of sensitive membrane proteins, **Immunol.**, 60:269.

Petty, H. R., Francis, J. W., and Boxer, L. A., 1988, Deficiency in immune complex uptake by chronic granulomatous disease neutrophils, **J. Cell. Sci.**, 90:425.

Pryor, W. A., 1981, Free radical biology: Xenobiotics, cancer and aging, **Ann. N.Y. Acad. Sci.**, 393:1.

Root, R. K., and Metcalf, J. A., 1977, H_2O_2 release from human granulocytes during phagocytosis: Relationship to superoxide anion formation and cellular catabolism of H_2O_2: Studies with normal and cytochalasin B treated cells. **J. Clin. Invest.**, 60:1266.

Schrachraufstatter, I., Revak, S. D., and Cochrane, C. G., 1984, Biochemical factors in pulmonary inflammatory disease, **Fed. Proc.**, 43:2807.

Shigeoka, A. O., Charette, R. P., Wyman, M. L., and Hill, H. R., 1981, Defective oxidative metabolic responses of neutrophils from stressed neonates, **J. Pediatr.**, 98:392.

Spielberg, S. P., Garrick, M. D., Corash, L. M., DeB Butler, J., Tietze F., Rogers, L., and Schulman, J. D., 1978, Biochemical heterogeneity in glutathione synthetase deficiency, **J. Clin. Invest.**, 61:1417.

Spielberg, S. P., Boxer, L. A., Oliver, J. M., Allen, J. M., and Schulman, J. D., 1979, Oxidative damage to neutrophils in glutathione synthetase deficiency, **Br. J. Haemotol.**, 42:215.

Stolc, V., 1979, Characterization of iodoproteins secreted by phagocytosing human polymorphonuclear leukocytes, **J. Biol. Chem.**, 254:1273.

Suzuki, T., Saito-Taki, T., Sadasivan, R., and Nitta, T., 1982, Biochemical signals transmitted by Fc receptors: Phospholipase A_2 activity of Fc2b receptors of murine macrophage cell line P 388D1, **Proc. Natl. Acad. Sci. USA**, 79:591.

Tauber, A. I., Borregaard, N., Simons, E., and Wright, J., 1983, Chronic granulomatous disease: A syndrome of phagocyte oxidase deficiencies, **Medicine**, 62:286.

Till, G. O., Johnson, K. J., Kunkel, R., and Ward, P. A., 1982, Intravascular activation of complement and acute lung injury: Dependency on neutrophil and toxic oxygen metabolites, **J. Clin. Invest.**, 69:1126.

Till, G. O., Hatherill, J. R., Tourtellotte, W. W., Lutz, M. J., and Ward, P. A., 1985, Lipid peroxidation and acute lung injury after thermal trauma to skin. Evidence of a role for hydroxyl radicals, **Am. J. Pathol.**, 119:376.

Thomas, E. L., 1979, Myeloperoxidase, hydrogen peroxide, chloride antimicrobial system: Nitrogen-chlorine derivatives of bacterial components in bactericidal action against *Escherichia coli* , **Infect. Immun.**, 23:522.

Young, J. D.E, Unkeless, J. C., Kabacks, H. R., and Cohn, Z. A., 1983, Mouse macrophage Fc receptor for IgG γ2b/γ, in artificial and plasma membrane vesicles functions as a ligand-dependent ionophore, **Proc. Natl. Acad. Sci. USA**, 80:2636.

Weiss, S. J., Lampert, M. B., Test, S. T., 1982, Chlorination of taurine by human neutrophils: Evidence for hypochlorous acid generation, **J. Clin. Invest.**, 70:598.

Wright, W. C., Jr., Ank, B. J., Herbert, J., and Stiehm, R., 1975, Decreased bactericidal activity of leukocytes of stressed newborn infants, **Pediatrics**, 56:579.

ANTIOXIDANT VITAMINS AND THEIR FUNCTIONS IN IMMUNE RESPONSES

Adrianne Bendich

Clinical Nutrition
Hoffmann-LaRoche Inc.
Nutley, NJ 07110

INTRODUCTION

Vitamin E, vitamin C and beta carotene (pro-vitamin A) are essential nutrients which cannot be synthesized in the human body. Each of these nutrients has its own profile of distinct functions. In addition to these unshared functions, these three micronutrients share the capacity to act as antioxidants. As discussed in the introductory chapter, an antioxidant has the ability to stabilize highly reactive, potentially harmful molecules called free radicals. The generation of free radicals has been associated with damage to membranes, enzymes as well as the cell's nuclear material. The antioxidant's ability to lower the burden of highly reactive free radicals serves to protect the structural integrity of cells and tissues of the immune system as well as other systems in the body (Machlin and Bendich, 1987).

The major emphasis of this chapter will be on the immunomodulating effects of the fat-soluble antioxidant vitamin, vitamin E and water-soluble, vitamin C, since we and others have found these vitamins are potent immunoenhancers when supplemental levels are added to diets of laboratory and farm animals, and when supplements are given to humans. Research on the immunomodulating effects of beta carotene in humans is preliminary, however, there are a number of reports of efficacy in animal models and these are reviewed.

VITAMIN E

The initiation of the immune response occurs at the cell membrane (Roitt, 1984). Vitamin E is an essential constituent of all the membranes found in cells including the plasma, mitochondrial and nuclear membranes (Machlin, 1984). If vitamin E is absent from the diet, one would predict damage to cell membranes including those of lymphocytes. In fact, electron micrographs of mitochondrial membranes from lymphocytes, reticulocytes and platelets from vitamin E deficient rats show swollen and disrupted areas (Lehmann and McGill, 1982). All circulating cells which are derived from the common hemopoietic stem cell precursor demonstrate membrane changes when vitamin E is absent from the diet. Red blood cells from vitamin E deficient animals and humans lyse in the presence of hydrogen peroxide whereas red blood cells from vitamin E supplemented animals are not lysed. There is a significant increase in platelet number in vitamin E deficient rats (Bendich et al., 1986a). Platelets from vitamin E deficient animals are more adhesive and produce more thromboxane than platelets from vitamin E-replete animals (Machlin, 1984).

The vitamin E content of lymphocytes and mononuclear cells is approximately ten times greater than that found in platelets and red blood cells (Table 1; Hatam and Kayden, 1979). When macrophages are exposed to oxidative stress, the vitamin E content of these cells is significantly reduced (Coquette et al., 1986).

Table 1. Vitamin E Content of Blood Components

Blood Component	Vitamin E (ug/10^9 cells)
Neutrophils	4.5
Lymphocytes	3.9
Monocytes	25[a]
Erythrocytes	0.2
Platelets	0.5
Plasma	8.8[a]

[a]Monocyte value is based upon resident peritoneal rat macrophage values (Coquette et al., 1986); plasma value is expressed in ug vitamin E/ ml plasma; all other values are from humans (Hatam and Kayden, 1979).

In vitro Studies

Cell culture conditions are conducive for the production of reactive oxygen intermediates. The oxygen tension is higher than that found in lymphoid organs and the oxidative products of cultured cells are not usually removed during the culture periods, which may be 48 - 72 hr. Free radical mediated reactions can result in membrane lipid peroxidation. Peroxidation of lymphocyte membranes significantly depresses in vitro responses to mitogens as well as primary antibody responses. Addition of vitamin E to lymphocyte cultures overcomes the immunosuppression (Hoffeld, 1981). Murine splenocytes exposed to silica and other small particles show depressed immune responses which were reversed when the cells were cultured with vitamin E (Hoffeld, 1983). In both studies, addition of the synthetic antioxidant, 2-mercaptoethanol, produced similar effects as vitamin E.

Macrophages and neutrophils have the capacity to take up molecular oxygen and generate reactive oxygen-containing molecules when stimulated. This is often called the oxidative burst. Free radicals and singlet oxygen, along with other reactive molecules, can kill bacterial pathogens. Neutrophils can also generate highly toxic halogenated molecules when the myeloperoxidase halide enzyme system is activated during the oxidative burst. The halogenated species can also lyse pathogens. The killing process is usually confined to intracellular vacuoles which enclose the phagocytized pathogen.

Neutrophils from vitamin E deficient animals have an increased level of peroxidized lipid in their membranes, produce more hydrogen peroxide, and have depressed chemotaxis and phagocytosis (Machlin, 1984). Vitamin E deficient rats have impaired neutrophil and macrophage chemotaxis, reduced ingestion of complement coated beads and decreased protection from autooxidative damage (Harris et al. 1980). At birth, premature infants have a low vitamin E status and poor neutrophil functions. Following intramuscular injections of vitamin E, in vitro measurements of chemotaxis and phagocytosis during the first five days of life reached the level found in normal term infants and were significantly enhanced over the level seen in unsupplemented premature infants (Chirico et al., 1983).

Chemiluminescence is often used as an index of oxygen burst activity. In vitro measurement of macrophages from vitamin E-deficient rats showed a threefold increase in chemiluminescence compared to macrophages from rats fed vitamin E. Incubation of the deficient macrophages with vitamin E decreased the degree of chemiluminescence. There was also an increase in the viscosity of the vitamin E-deficient macrophage membranes. (Sharmanov et al., 1986). Vitamin E is important in maintaining lymphocyte membrane fluidity necessary for proliferative responses to mitogens (Fountain et al., 1982).

Animal Studies

The initial studies on the effect of vitamin E on immunity utilized higher than normal dietary levels fed to laboratory and farm animals. The animals were then challenged with pathogens. Responses were measured in terms of increased resistance, decreased recovery or decreased mortality (Table 2). In almost every instance, supplemental vitamin E enhanced immune responses. Antibody titers and phagocytosis of the pathogens were increased, and in several instances morbidity and mortality from infection were diminished. Neutrophil-induced oxidative damage to lung tissue following burn injury was significantly reduced when rats were preinjected with vitamin E (Till et al. 1985). In contrast to *in vitro* studies cited above, substitution of a synthetic antioxidant for vitamin E in the diet did not result in immunoenhancement. These findings have been reviewed and summarized recently (Tengerdy et al. 1984; Tengerdy, 1989; Chow, 1985; Panush and Delafuente, 1985; Bendich, 1988a).

When laboratory animals are placed on vitamin E deficient diets, T and B lymphocyte responses to mitogenic stimulation (Bendich et al. 1986a), mixed lymphocyte responses (Corwin and Gordon, 1982) and plaque forming cell levels were severely depressed (Corwin and Schloss, 1980a; Gebremichael et al. 1984). Macrophage membrane receptors for Ia antigens were decreased (Gebremichael et al. 1984), interleukin 2 production was lowered (Bendich, 1988a) and production of immunosuppressive prostaglandins were decreased (Likoff et al. 1981; Meydani et al. 1984, 1986; Lawrence et al. 1985).

When lymphocytes were cultured with a synthetic antioxidant, such as 2-mercaptoethanol, the lymphocytes from vitamin E deficient rats and mice did not proliferate when stimulated with mitogens to the level found with lymphocytes from animals fed vitamin E (Bendich et al., 1986a). This would imply that the vitamin E must be present during the synthesis of all cellular components. Low levels of dietary vitamin E (7.5 and 15 mg/kg in the diet) were insufficient to enhance T and B lymphocyte mitogenic responses even though the purified diet contained all other nutrients at the level recommended by the American Institute of Nutrition (Fig 1, Bendich et al., 1986a).

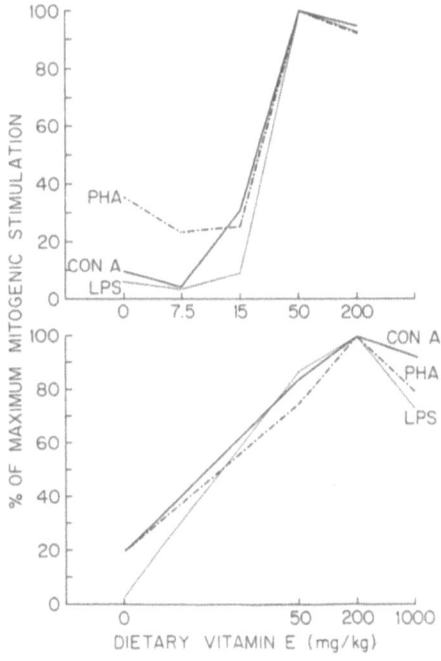

Figure1. The effect of dietary vitamin E on lymphocyte proliferation.

Table 2. Vitamin E Supplementation and Enhanced Responses to Infective Agents.

Animal	Dosage	Infectious Organism	Effects	Reference
Chicks	100 IU/kg diet	*Eimeria tenella*	-decreased mortality -increased weight gain	Colnago et al., 1984
Chicks	150 or 300 IU/kg diet	*E. coli*	-decreased mortality (both levels)	Heinzerling et al., 1974b
Chicks	150 or 300 IU/kg diet	*E. coli*	-decreased mortality (both levels) -increased resistance	Nockels 1979
Chicks	300 IU/kg diet	*E. coli*	-decreased mortality -enhanced recovery	Tengerdy & Nockels, 1975
Turkeys	100 or 300 IU/kg diet	*E. coli*	-decreased mortality (both levels) -increased resistance	Nockels 1979
Turkeys	500 IU/kg diet and	*Histamonas meleacridis*	-decreased mortality and lesion scores	Schildknecht & Squibb, 1979
Sheep	1,000 IU/day 1, plus 300 IU/kg diet	*Chlamydia*	-increased resistance	Nockels 1979
Lambs	1,000 IU/day (oral)	*Chlamydia*	-enhanced recovery	Stephens et al., 1979
Calves	1800 IU/wk diet (oral) 1400 IU/wk diet (injection)	Rhinotracheitis virus	-inhibited viral replication	Reddy et al., 1965
Pigs	20,000 IU/ton diet 100,000 IU/ton diet	*E. coli*	-increased resistance	Ellis & Vorhies, 1976
Pigs	200 IU/day	*Treponema hyodysenteriae*	-increased resistance	Teige et al., 1982
Mice	60 IU/kg diet 120 IU/kg diet	Sheep red blood cells Tetanus toxoid	-improved humoral immune response	Nockels 1979
Mice	180 IU/kg diet	*Diplococcus pneumoniae*	-decreased mortality	Heinzerling et al., 1974a
Rats	180 IU/kg diet and 600 IU vit. A/kg diet	*Mycoplasma pulmonis*	-increased resistance	Tvedten et al., 1973

With fat soluble vitamins, a tenfold increase in dietary concentration may be required to achieve a twofold increase in serum vitamin levels. There is also individual variability in absorption, metabolism and excretion. In the above experiments, the log of the level of vitamin E in the diet was significantly correlated with the plasma vitamin E levels. More importantly, plasma vitamin E levels were significantly correlated with optimum responses to T and B lymphocyte mitogen responses (Bendich et al., 1986a). In agreement with this finding of a correlation of high plasma vitamin E levels with enhanced lymphocyte responses, Chavance et al. (1984) found a significant association between high plasma vitamin E status (> 1.35 mg/dl) and a lower number of infections in a healthy human population over the age of 60. Mbawuike et al. (1982) found a positive correlation between splenic vitamin E level and mitogenic activity of splenocytes in mice.

In certain instances, the immune system can develop aberrant responses which can result in harmful self-destructive responses. An example would be the chronic inflammatory response seen in the joints of arthritics. The damage caused in these joints has been associated with the increase in free radicals generated in the inflamed tissues. The level of dietary vitamin E has been directly linked to the degree of *in vivo* lipid peroxidation and inflammatory responses. Vitamin E deficient rats given a bolus injection of 100 mg vitamin E/gm body weight had significantly less inflammation (as measured by less swelling of the foot pad following injection of Freund's complete adjuvant) and lower expiration of pentane (index of lipid peroxidation) than vitamin E deficient rats. Furthermore, vitamin E protected against the chronic inflammation and bone distortion (Dillard et al. 1982).

In contrast to the consensus of data indicating the requirement for vitamin E for optimal T and B lymphocyte and macrophage function, splenic natural killer cell ability to lyse tumor cells was not altered significantly when either mice (Kurek and Corwin, 1982; Meeker et al., 1985) or rats (Bendich et al. 1985) were fed vitamin E deficient or supplemented diets. However, when tumor cells were injected into mice, tumors were more numerous and faster growing in the vitamin E deficient group (Kurek and Corwin, 1982). Since the mechanism of natural killer cell lysis does not seem to involve oxidative damage to the tumor cells (Herberman, 1985), further research into the protective role of vitamin E in tumor development is warranted.

It should be noted that as a dietary antioxidant, vitamin E protects other immunoenhancing substances such as vitamins C and A from oxidation. As the major vitamin antioxidant in plasma and tissues, it may also protect against free radical destruction of these and other nutrients (Alfin-Slater and Morris, 1963; Bendich et al. 1984, Wayner et al. 1985).

Vitamin E and Dietary Lipids

Because it is an antioxidant, the level of vitamin E required in the diet is dependent, in part, on the level of potentially oxidizable fatty acids in the diet (RDA,1980). High levels of dietary PUFA have been shown to be immunodepressing (Mertin and Hunt, 1976; Erickson et al., 1980; Newberne, 1981; Gurr, 1983). The mechanism of lipid-induced immunosuppression may be via an alteration in the fluidity of cell membranes which is, in part, dependent upon the degree of unsaturation of its fatty acids. As the level of PUFA is increased, the potential for lipid peroxidation is also increased. Lipid peroxidation causes a decrease in membrane fluidity (Meade and Mertin, 1978). In addition to consequent loss in fluidity, metabolites of lipid peroxidation can adversely affect immune responses (Mertin and Hughes, 1975; Erickson et al. 1983). Loss of membrane fluidity has been directly related to the decreased ability of lymphocytes to undergo mitogenesis (Fountain and Schultz, 1982).

If the fat source in animal diets is corn oil (high in polyunsaturated fatty acids (PUFA)) rather than lard or coconut oil (high in saturated fatty acids) immune responses of T and B lymphocytes to mitogens and *in vitro* antigen dependent antibody production are depressed. The degree of immunodepression is directly related to the concentration of dietary PUFA. Shapiro et al. (1984) found no depression of mouse T and B lymphocyte mitogen responses when 5% PUFA replaced 5% saturated fat in the diet. Corwin and Schloss (1980b) found depression when 8% PUFA replaced saturated fat in mouse diets. Bendich et al. (1988a) found significant depression in mitogen responses when rats were fed 10% PUFA instead of lard. Vitamin E deficiency as well as low vitamin E exacerbated this immunosuppression. Higher than normal dietary levels of vitamin E could partially overcome the immunosuppressive effects of high PUFA diets.

Vitamin E and Arachidonic Acid Breakdown

A primary PUFA in lymphoid cell membranes is arachidonic acid. Activation of the membrane-bound enzyme, phospholipase A_2 releases arachidonic acid from the membrane phospholipid, exposing arachidonic acid to degradation by the endoperoxidases cyclooxygenase or lipoxygenase. The products of cyclooxygenase activity include prostaglandin E_2, which is immunosuppressive. Lipoxygenase products include several chemotactic leukotrienes. The formation of prostaglandins and leukotrienes involve the uptake of oxygen and is an important example of enzyme-catalyzed lipid peroxidation.

As the major lipid-soluble membrane antioxidant, vitamin E, has been shown to downregulate many of the pathways involved in the metabolism of arachidonic acid. Douglas et al. (1986) demonstrated that vitamin E deficient rats had enhanced platelet phospholipase A_2 activity. The enzyme activity was lowered in a concentration-dependent manner as the level of vitamin E in the diet or test tube was increased. Meydani et al. (1986) showed that high levels of dietary vitamin E decreased the level of splenocyte prostaglandin E_2 and enhanced lymphocyte functions. In addition, vitamin E has been shown to directly inhibit the 5-lipoxygenase enzyme responsible for the formation of leukotrienes (Reddanna et al., 1985). Although all of these studies were not evaluating the effects of arachidonic acid metabolism on aspects of immune function, there is strong evidence that alteration of arachidonic acid degradation is one of the prime mechanisms by which vitamin E affects immune responses.

Vitamin E, Prostaglandins and Immune Functions

As the level of PUFA in the diet is increased, the level of unsaturated lipids, including arachidonic acid, in membrane phospholipids also increases. As discussed above, vitamin E can alter arachidonate metabolism and consequently, prostaglandin (PG) levels. Corwin and Schloss (1980a) found that vitamin E added to diet or cultures of mouse lymphocytes did not affect PG production but did enhance mitogenesis whereas indomethacin (a cyclooxygenase inhibitor) added to cultures depressed PG production and did not enhance mitogenesis. In contrast to these findings, Deshago et al. (1981) found that both indomethacin and vitamin E (singly or together) enhanced the mitogenesis of lymphocytes from Hodgkin's disease patients. Although these investigators did not measure PG levels, Likoff et al. (1981) found that 300 mg/kg vitamin E added to control and *E. coli* infected chicken diets decreased PGE_2 levels significantly in spleen and bursal tissues. When aspirin (also a cyclooxygenase inhibitor) was injected into chicks fed the vitamin E supplemented diet, these two substances acted synergistically to protect chicks from death caused by *E. coli* infection. Recently, Meydani et al. (1986) have reported that 500 mg/kg vitamin E added to diets of 24 month old mice (aged) restored the immunosuppressed responses of splenocytes to mitogen stimulation as well as *in vivo* cell mediated immune responses to a contact sensitizer to the level found in 4 month old mice. These enhanced responses were correlated with a depression in splenic PGE_2 production in the older mice given vitamin E.

Vitamin E and Selenium

Along with the antioxidant vitamins, there are several enzymes which can protect against damage caused by free radicals. While vitamin E can protect against lipid peroxidation in cell membranes, the enzyme glutathione peroxidase can reduce lipid peroxides within the cell. Selenium is required for the activation of the metalloenzyme glutathione peroxidase. The enzyme is also involved in the destruction of hydrogen peroxide (Shamberger, 1983).

Both vitamin E and selenium are required for optimal immune function. Several investigators have examined the potential for either essential nutrient to substitute for the antioxidant effects of the other. Bendich et al. (1988a) have found that selenium cannot substitute for vitamin E and enhance T and B lymphocyte responses to optimal levels. Vitamin E supplementation of selenium deficient diets resulted in enhancement of mitogen responses. The highest responses were in groups fed both nutrients. Using a similar 2x2

40

factorial design, Eskew et al. (1985) also found that vitamin E and selenium deficient rats had severely depressed mitogen responses and addition of both selenium and vitamin E enhanced T lymphocyte responses significantly. B lymphocyte responses were also depressed when both nutrients were absent from the diet.

Sheffy and Schultz (1979) have reported that dietary selenium could not replace dietary vitamin E to reverse the immunosuppression caused by selenium and vitamin E deficiency in dogs. In an experiment reported by Mulhern et al. (1985) the lymphocyte mitogen responses of first generation selenium deficient, vitamin E adequate mice were the same as animals fed the diet containing selenium , however specific antibody responses (IgG) were reduced following challenge with sheep red blood cells. When mice were kept selenium deficient through two generations, mitogenesis, specific and non specific antibody titers as well as thymic size were depressed even though mice were fed adequate levels of vitamin E (Mulhern et al. 1985).

Larsen and Tollersrud (1981) reported that pigs fed a vitamin E and selenium deficient diet for 12 weeks had significantly depressed mitogenic responses to PHA when compared to pigs fed the double supplemented diet. Addition of selenium alone enhanced these responses somewhat however, addition of vitamin E significantly increased these responses irrespective of the levels of selenium used.

Vitamin E and selenium deficient mice had depressed natural killer cell lysis of tumor target cells. When vitamin E was added to the diet, these responses were restored to the level seen in mice fed both nutrients. Antibody dependent cell mediated cytotoxic responses were unaffected by dietary levels of these nutrients (Meeker et al., 1985).

Although both nutrients appear to be required for optimal immune responses, in many experiments, vitamin E is the more critical agent.

Tumor Models

Several investigators have shown an association of vitamin E deficiency with increased growth of tumors in animal models (Cook and McNamara, 1980). Some studies have shown a decrease in tumor burden with supplementary vitamin E (Kurek and Corwin, 1982) whereas other studies have not replicated the positive findings (Wattenberg, 1972). It has been only recently that investigators have looked at the effect of vitamin E on tumor immune responses. The capacity of vitamin E to enhance immunity to tumors was evaluated in a carcinoma model using hamsters. Injections of vitamin E into the tumor-bearing cheek pouch resulted in a regression of the tumor (Shklar et al., 1987). The concentration of tumor necrosis factor, secreted by activated macrophages, was significantly enhanced in the vitamin E-treated group (Shklar and Schwartz, 1988).

Human Studies: Status

The relationship between vitamin status and immunological status was assessed in 100 healthy subjects over 60 years old. Chavance et al. (1984) found that subjects with serum vitamin E levels > 1.35 mg/dl had significantly lower numbers of infections over a three-year time span. Average serum vitamin E levels in the U.S. are 0.9-1.0 mg/dl.

Human Studies: Intervention

In a preliminary report, children with an increased susceptibility to respiratory tract infections and an OKT 4/8 ratio of 1/2 (helper/suppressor T lymphocyte ratio) were given 20 mg vitamin E/kg body wt/day for six weeks. The percent of OKT4 lymphocytes increased significantly and the ratio was normalized to 2/1. No functional assays of immune responses were measured. Of the eight children in the study, six remained healthy during the vitamin administration (Skopinska-Rozewska et al., 1987). Hemodialysis patients given intramuscular injections of 300 mg vitamin E for 15 days showed a significant increase in the OKT 4/8 ratio from 1.6 to 2.3. No effect on natural killer cell activity or mitogen responses were seen (Taccone-Gallucci et al., 1986).

There are seven studies on the effect of oral administration of vitamin E on immune responses in free living, healthy adults (Table 3). Five of the seven studies involve small numbers of individuals and most are uncontrolled. Most of the studies characterize the effect of vitamin E on neutrophil functions and three papers discuss the effects of supplementation on specific immune responses.

In the first double-blind, placebo controlled study to evaluate the effect of vitamin E on phagocytosis, Hill et al. (1983) showed that the depressed chemotaxis and random motility of

Table 3. Vitamin E and Human Immune Responses

Number of Individuals	Dosage*	Findings	Ref.
3	1600 mg/day/one week	a. neutrophil phagocytosis increased b. neutrophil bactericidal activity decreased c. auto-oxidative damage to neutrophils decreased	Baehner et al., 1977
3	1600 mg/day/two weeks	hydrogen peroxide level decreased in neutrophils	Butterick et al., 1983
1	400 mg/day/3 months	normalization of neutrophil functions; no new bacterial infections	Boxer et al., 1979
9	300 mg/day/3 weeks	a. no change in DTH to PHA b. decrease in mitogen response to PHA c. decrease in bactericidal activity	Prasad, 1980
4**	1200 mg/day/3 weeks	no change in MLC with vitamin E supplemented responder lymphocytes; decrease with vitamin E supplemented stimulators	Davey & Dock, 1982
14	25 mg/kg body wt/ day (approx 1500 mg/day) 2-3 weeks placebo-controlled, double blind study	normalization of depressed monocyte chemotaxis in diabetic subjects	Hill et al., 1983
34	800 mg/day/one month; placebo-controlled double blind study.	a. significantly enhanced delayed hypersensitivity responses in vitamin E group b. increase in mitogen response to Con A	Meydani et al., 1988

*RDA for vitamin E is 30 mg/day for adults.

**Four individuals were given the supplement, two individuals were used as matched controls. In all studies other than Meydani et al., (1988), there were no matched placebo-control groups.

monocytes from diabetic individuals were normalized following 2-3 wk of vitamin E supplementation with 25 IU/kg body wt/day (approximately 1500 IU/day).

Recently, Meydani et al. (1988; 1989) completed the first double-blind, placebo controlled study examining the effects of vitamin E supplementation on the immune responses in a healthy elderly population. Supplementation with 800 IU/day for approximately one month significantly enhanced delayed hypersensitivity responses (a significant indicator of ability to combat infectious disease) while the level of circulating lipid peroxides were diminished. Lymphocyte proliferation responses to concanavalin A were also significantly enhanced. No adverse side effects were noted in this well controlled study. The safety of high doses of vitamin E has recently been reviewed (Bendich and Machlin, 1988).

The ability of elderly to resist infections decreases with increasing age. These data suggest that elderly supplemented with vitamin E may be better able to resist infections than the control group. These data also provide a rational for the epidemiological study cited above in which elderly with high serum vitamin E levels had a lower incidence of infections (Chavance, 1984).

VITAMIN C

Vitamin C (ascorbic acid) has a number of biochemical functions which are linked to the functions of the immune system. These functions include the acceleration of hydroxylation reactions required for the formation of hydroxyproline from proline, enhancing the formation of collagen (Levine, 1986). Collagen is an essential component of basement membranes which serve as the attachment for epithelial cells lining the digestive, respiratory, urinary and reproductive tracts. In addition, collagen is an important component of skin and bone tissue. Vitamin C is therefore involved in the maintenance of the natural barrier.

Vitamin C is an important component of the overall antioxidant defense system which functions to protect against damage from free radicals. Singlet oxygen, although not a free radical, can participate in the formation of free radicals. Vitamin C can quench singlet oxygen (Bendich et al., 1986b). Vitamin C can block the formation of carcinogenic nitrosamines in the test tube as well as in the digestive tract (Ohshima and Bartsch, 1981).

Most of the experiments described above involve the effects of vitamin E on immune responses in mice and rats. For rodents and most animals, vitamin C is not a vitamin since it can be synthesized in their livers. Unlike these animals, man does not have the ability to synthesize vitamin C. This is also true for guinea pigs and monkeys (Burns, 1959). Therefore, studies using animals with a vitamin C requirement are emphasized.

In Vitro Studies

Within the blood, the circulating immunoglobulins (sometimes referred to as antibodies) and complement factors protect against systemic infections. Vitamin C has been shown to be required for the secretion of substances, such as immunoglobulins and interferons (Morre, 1987). Recent data have shown that circulating levels of the hydroxyproline-rich complement component, C1q paralleled dietary vitamin C levels in guinea pigs (Johnston et al., 1985).

Interferons are secreted by cells in response to viral infection. There are data from *in vitro* and *in vivo* experiments in mice which have shown that vitamin C enhances interferon levels (Siegel, 1974; Anderson, 1982).

Vitamin C has also been shown to be important for optimal phagocytosis. In fact, the concentration of vitamin C in neutrophils and macrophages is approximately 150 times the concentration of vitamin C in the plasma (Anderson, 1982; Castelli et al., 1982; Evans et al., 1982; Moser and Weber, 1983). During the oxidative burst, neutrophils take up vitamin C (Moser and Weber, 1983) and following activation, the vitamin C concentration is reduced (Hemila et al., 1985, Oberritter et al., 1986). Neutrophils from vitamin C deficient guinea pigs have depressed chemotaxis and bactericidal activities (Shilotri, 1977; Goldschmidt et al., 1988). Vitamin C supplementation enhances the chemotactic response of neutrophils from healthy individuals (Anderson, 1982; Beisel, 1982; Panush and Delafuente, 1985). Vitamin C

43

supplementation has also been found to normalize the reduced chemotactic and bactericidal activities of neutrophils from individuals with inherited phagocytosis disorders (Weening et al., 1981, Gallin, 1981) as well as from newborn infants (Vohra et al., 1983). In several instances, clinical improvements were found.

During the oxidative burst, reactive molecules are released into the surrounding area and can cause mutations, lyse normal cells including other neutrophils, cause inflammation of surrounding tissues, inactivate protective enzymes and inhibit lymphocyte proliferation. Anderson and Lukey (1987) have shown that vitamin C can decrease the damaging effects of the products of the oxidative burst without decreasing the intracellular concentration of reactive, bactericidal molecules (Table 4). In addition, vitamin C protected alpha 1 protease inhibitor (a protective enzyme found in the lung) from inactivation by the free radicals generated during the oxidative burst (Theron and Anderson, 1985; 1987). The mechanism for alteration of human phagocytic activity may also be due to ascorbic acid decreasing lipoxygenase activity (Schmidt et al., 1988).

Table 4. Vitamin C and Nonspecific Immunity

Target	Effect
skin	required for collagen synthesis
linings of mouth, gut	maintains basement membrane
interferon	important in production and secretion
complement components	required for the synthesis of C1q
phagocytic cells	enhances chemotaxis, protects cells from free radical damage

Animal Studies

When lymphocytes from vitamin C deficient monkeys were cultured *in vitro*, the lymphocyte proliferative responses were significantly diminished (Hsu, 1977), however these responses were unaffected in guinea pigs in two studies (Zweiman et al., 1966, Bendich et al., 1984) and depressed in one study (Thurman and Goldstein, 1979). Vitamin C deficient guinea pigs had depressed cytotoxic T lymphocyte killing of tumor cells (Anthony et al., 1979). Antibody production was lower in scorbutic (vitamin C deficient) guinea pigs in one study (Thurman and Goldstein, 1979) and unaffected in another study (Kumar and Axelrod, 1969; McMurray, 1984).

There are consistent adverse effects of vitamin C deficiency on the cell mediated immune responses of scorbutic animals. Specifically, there was a depression of responses to vaccination (Mueller and Kies, 1962), and delayed hypersensitivity responses (Zweiman et al., 1966), both indicators of memory function. Responses to foreign tissues were also significantly diminished, resulting in the increased survival of grafted skin (Kalden and Guthy, 1972) and decreased autoimmune responses (Mueller et al., 1962).

Table 5. Vitamin C and Specific Immune Responses

	Deficiency	Supplementation
Lymphocyte Proliferation	Decreased	Increased
Cytotoxicity	Decreased	Increased
Antibody Titers	No effect	Increased, No effect
Delayed Hypersensitivity	Decreased	Increased
Response to Vaccines	Decreased	ND[a]
Inflammatory Responses	Decreased	ND
Autoimmune Responses	Decreased	ND

[a] ND = no data available

Vitamin C and Vitamin E Interactions

Vitamin C supplementation may indirectly improve immune responses by helping to maintain circulating and tissue levels of vitamin E (Bendich et al., 1984). The reduced form of vitamin C, ascorbic acid, can donate an electron to the tocopheryl free radical, regenerating the antioxidant form of vitamin E (Bendich et al., 1986b). Experiments to determine the interaction of these two vitamins on T and B lymphocyte responses to mitogens were undertaken in guinea pigs fed diets with either a low, but anti-scorbutic level of vitamin C or a high level of vitamin C and either a deficient, standard or supplemental level of dietary vitamin E. As in mice and rats, vitamin E deficient guinea pigs had depressed responses to mitogens compared to guinea pigs fed diets containing vitamin E. The two levels of vitamin C did not affect these responses (Bendich et al. 1984). In a related series of experiments, however, when guinea pigs fed the same diets were exposed to 100 % oxygen (with a high potential of generating free radicals and other energized molecules), those animals fed the vitamin E deficient diet and high vitamin C had enhanced T and B mitogen responses compared to the group fed the low vitamin C diet. Importantly, animals fed the high levels of both vitamins had the best overall responses including weight gain (Bendich et al. 1983). In both the 100 % oxygen exposed as well as the unexposed animals the level of lung vitamin E was highest in animals fed the high vitamin C diet. Plasma vitamin E levels were significantly higher in animals fed high dietary vitamin C.

The supplemental level of vitamin C protected tissue levels of vitamin E. Guinea pigs fed diets containing low vitamin C and the standard level of vitamin E had approximately half the concentration of vitamin E in their lung tissue as guinea pigs fed the same amount of vitamin E and the high concentration of vitamin C (Bendich et al., 1984). Kanazawa et al. (1981) and Hruba et al. (1982) have also shown that chronic vitamin C deficiency lowered the vitamin E levels in tissues and plasma.

Tumor Models

Vitamin C supplementation has enhanced survival, reduced tumor burden and growth in several animal models. There are also studies where vitamin C was not effective in tumor models (Glatthaar et al., 1986). There are no studies which have examined the ability of vitamin C to enhance tumor immunity specifically, although the enhancement of interferon production and T cell activities should encourage such experiments.

Human Studies

Several investigators have found that vitamin C supplementation (at 1 gm/day) enhanced human *in vitro* lymphocyte proliferative responses and antibody responses (Panush et al., 1982; Beisel, 1982; Anderson, 1982). In on study, supplementation did not enhance antibody titers but enhanced prealbumin levels (Dobias et al., 1986). Delayed hypersensitivity responses were also enhanced following oral supplementation (10 gm/day) in one study (Panush et al., 1982) in young adults and following injections of 500 mg/day of vitamin C in another study in an elderly population (Kennes et al., 1983). Oral supplementation (2 gm/day) in an elderly population enhanced *in vitro* lymphocyte proliferative responses but did not affect delayed hypersensitivity responses in another study (Delafuente et al., 1986).

BETA CAROTENE

Beta carotene is best known as the precursor of vitamin A found in fruits and vegetables. However, beta carotene is also a potent quencher of singlet oxygen and an antioxidant (Burton and Ingold, 1984).

Because of its ability to act as both a quencher of singlet oxygen and free radicals and its epidemiological association with lower risk of certain cancers, there are several groups studying the possible role of beta carotene as an immunomodulator (Bendich, 1988b).

Several complications arise in research with beta carotene. It is highly insoluble, making *in vitro* experiments difficult. Studies using rodents are also difficult since they are very efficient convertors of beta carotene to vitamin A, therefore making it questionable whether any immunoenhancement seen is due to the known immunostimulatory capability of vitamin A. There are, however, other carotenoids, such as canthaxanthin and astaxanthin, with chemical structures very similar to beta carotene. Canthaxanthin can serve as an appropriate dietary control since it is a singlet oxygen quencher but it cannot be converted to vitamin A.

In Vitro Studies

Although the formation of solutions with beta carotene is difficult, some *in vitro* studies have been reported. Studies by Rhodes (1983; Rhodes et al., 1984) have shown that beta carotene blocks the inhibiting effects of vitamin A on *in vitro* interferon-induced human monocyte receptor expression. Using a lymphoblastoid cell proliferation assay stimulated with interferon, beta carotene was shown to reverse the immunosuppressive effects of retinol. Tumors can produce molecules that suppress certain macrophage functions. Beta carotene was also found to overcome the inhibition of interferon-induced macrophage activation by suppressor molecules synthesized by tumors (Gruner et al., 1986).

Interferons have many immunological functions, among which are stimulation of natural killer cell cytotoxicity. In a preliminary report (Leslie and Dubey, 1982), when alpha or beta carotene was added to cultures of human lymphocytes, there was an increase in natural killer cell lysis of tumor cells.

Animal Studies

Seifter et al. (1981) examined the effects of short-term intake of beta carotene in mice. They found an increase in thymic weight, stimulation of allograft rejection and inhibition of virally induced tumor growth. This group used bixin, another carotenoid which is not a precursor of vitamin A and has a more marked difference in chemical structure than canthaxanthin, as a dietary control. They found no effect of bixin on the immune parameters measured. Comparisons were also made with citral and *all-trans* retinoic acid, both of which depressed thymic weight and inhibited allograft rejection.

46

Both dietary beta carotene and canthaxanthin enhanced rat T and B lymphocyte proliferative responses (Bendich and Shapiro, 1986). In these experiments, mitogen responses were assessed with both carotenoids and then compared with plasma and tissue vitamin A, beta carotene and canthaxanthin levels. The results point to an immunoenhancing capacity of beta carotene separate from its pro-vitamin A activity which is hypothesized to be linked to the antioxidant function shared by both carotenoids.

Tumor Models

Epidemiological studies point to a strong chemoprotective role for carotenoids. The mechanism for this protection has not as yet been elucidated. Recently, several laboratories have examined the ability of beta carotene and other carotenoids to enhance the cell-mediated immune responses to tumors in animal models, resulting in tumor regression and/or reduced growth of tumors.

Injection of murine sarcoma virus into mice resulted in the formation of tumors. Supplementation of standard diets with beta carotene for three days prior to and continuing after viral inoculation was associated with an increase in the latency period, slowing of tumor growth and an increase in the rate of tumor regression. In the same experiment, bixin, a carotenoid lacking provitamin A activity and having antioxidant capacity increased the latency period but had less of an effect on tumor growth and regression than beta carotene. Citral, isolated from lemons (which is not a carotenoid or retinoid) and retinoic acid enhanced tumor growth (Seifter et al., 1982). Retinol (vitamin A) was not tested in this model, however it has been shown to protect mice from viral sarcoma tumor development (Seifter et al., 1981). Therefore, beta carotene may be serving as a source of vitamin A in this experiment.

Adenocarcinoma cells were subcutaneously injected into healthy mice. When the tumors had reached a palpable size, the mice were divided into three dietary groups. The control group was fed a lab chow containing standard levels of vitamin A and beta carotene. One experimental group was fed the chow diet supplemented with a high level of vitamin A and the other experimental group was fed the chow diet with the same level of retinol equivalents in the form of beta carotene. Both beta carotene and vitamin A supplemented groups had slower growing tumors than the control group (Seifter et al., 1983; 1984).

A subcutaneous injection of carcinoma cells was administered to the cheek pouch of hamsters. Local injection of beta carotene into the same cheek pouch significantly lowered the number of tumors which developed/cheek pouch as well as decreased the incidence of tumors. Macrophages taken from tumor bearing hamsters treated with beta carotene killed more of the tumor cells *in vitro* than macrophages from tumor bearing controls. In addition, the ability of the macrophages to generate tumor necrosis factor (TNF) was enhanced significantly in the beta carotene treated group (Schwartz et al., 1986). Administration of canthaxanthin resulted in similar decreases in tumor burden and increases in TNF whereas 13-cis retinoic acid administration increased tumor growth (Shklar and Schwartz, 1988).

Fibrosarcoma cells were subcutaneously injected into mice which were then given orally administered beta carotene or placebo for nine days. The primary tumor was excised and a second injection of the same tumor or a different tumor was given. Secondary challenge of the immune response to the initial tumor was enhanced significantly in the beta carotene group. The tumors were half the size of those in the placebo group. The groups given beta carotene and injected with a different tumor the second time had tumors which were the same size as those found in the placebo group (Tomita et al., 1987a).

In an extension of this protocol, tumor bearing mice, either given beta carotene, canthaxanthin or astaxanthin or placebo, were sacrificed. Lymph-node cells were removed and then mixed with fresh tumor cells. The cell mixture was then inoculated into healthy mice. Those mice receiving lymphnode cells from carotenoid-treated mice had one seventh the tumor burden of mice given lymphnode cells from placebo treated mice. If the lymphnode cells from the carotene groups were exposed to a substance that destroyed cytotoxic T lymphocyte activity, and then injected with tumor cells into healthy mice, the tumor growth was no longer curtailed (Tomita et al., 1987b).

The application of 7, 12-dimethylbenz(a)anthracene (DMBA) to the hamster cheek pouch for 22 wk caused the formation of tumors. When beta carotene was applied to these tumors for 4 wk, there was a significant decrease in the number of tumors found and 8/10 hamsters had no tumors in the cheek pouch. All hamsters in the placebo group (n=10) had tumors present at the end of this experiment. Twice as many macrophages from the beta carotene-treated cheek pouches contained tumor necrosis factor (Schwartz et al., 1986).

Table 6. Carotenoids Examined in Tumor Models

Carotenoid	Tumor type/source	Animal model
beta carotene bixin[a]	virally-induced sarcoma	mouse
beta carotene	adenocarcinoma cells	mouse
beta carotene	squamous cell carcinoma cells	hamster
beta carotene canthaxanthin astaxanthin	fibrosarcoma cells	mouse
beta carotene canthaxanthin phycotene	DMBA-induced oral carcinoma	hamster
beta carotene	DMH-induced adenomas and adenocarcinomas	mouse

[a]Other than bixin, all other carotenoids show activity in the tumor models examined.

In a second series of experiments, DMBA was applied for 13 wk to the hamster cheek pouch. Beta carotene, canthaxanthin, phycotene (algal extract) or 13-cis retinoic acid was injected into the DMBA-treated cheek pouch twice weekly for 4 wk. Following treatment with beta carotene, canthaxanthin or phycotene, complete regression of tumors were found in 4, 3, and 6/20 hamsters respectively. Partial regression of tumors were noted in all remaining animals in these three groups as well as in 14/20 in the 13-cis retinoic acid-treated group (Schwartz and Shklar, 1987).

In the most recent studies by this group, one fifth the concentration of DMBA was applied to the cheek pouch for 28 wk. In contrast to the above experiments in which carotenoids were given following the development of the tumor, in this study, beta carotene, canthaxanthin, or phycotene were orally administered during the time of DMBA treatment. Tumor formation was blocked completely by canthaxanthin and significantly decreased by the other treatments. Both macrophage and T lymphocyte cytotoxic activities were significantly increased and tumor necrosis factor levels were enhanced by all treatments compared to control levels (Schwartz and Shklar, 1989).

Injections of 1,2 dimethylhydrazine (DMH) resulted in the formation of tumors in the colon of mice following 31 wk of treatment. Mice fed a diet containing a low concentration of beta carotene for five weeks prior to DMH treatment and continuing until the time of sacrifice had significantly fewer tumors and fewer mice developed tumors compared to controls. Specifically, there was a 40 % reduction in the incidence and number of adenomas and a 96 % reduction in the incidence and 86 % reduction in the number of adenocarcinomas (Temple and Basu, 1987).

A number of the studies discussed above have demonstrated that carotenoids lacking vitamin A activity can significantly decrease the number and rate of growth of tumors through the enhancement of tumor immunity. Other factors may be found to be involved in the depression of tumor growth, however the finding by both groups of enhanced cytotoxic T lymphocyte activity with carotenoid administration should encourage increased research on the immunoenhancing properties of carotenoids.

Human Studies

There is only one experiment in the literature today on the effects of administration of beta carotene on immune parameters in humans (Alexander et al., 1985). Young men, who were supplemented with 180 mg/day of beta carotene, had an increase in the number of circulating T helper cells. Since T helper cells are involved in the regulation of the immune response, this is an interesting finding which should be further explored.

CONCLUSIONS

The three micronutrients reviewed in this chapter are unique in being the only dietary sources of direct acting antioxidants. Vitamin E, vitamin C and beta carotene supplementation to the diets of several animal species has resulted in enhancement of lymphocyte proliferative responses as well as other measures of host immunity. Consistent findings of depressed immune responses (nonspecific and specific, involving neutrophils, T and B lymphocytes, macrophages and lymphokine production) in vitamin C and E-deficient animals points to the importance of these essential nutrients for optimal immune functions. Beta carotene deficiency experiments would require concomitant vitamin A deficiency, which is known to be immunodepressive.

All three micronutrients have been shown to reduce and/or delay the growth of tumors in animal models. With vitamin E and beta carotene, there is evidence of enhanced tumor immunity. There is also epidemiological data associating diets low in any of the antioxidant vitamins and/or beta carotene with increased risk of cancer.

Future research associating vitamin C, vitamin E and beta carotene with enhanced protection from infective agents, decreased tumor growth and/or regression, increased vaccination responses, abolition of radiation-induced immunosuppression and improvement in wound healing are warranted.

It is hoped that this review will stimulate further research in animal models and humans with the individual vitamins as well as with combinations of these unique, safe micronutrients. The safety of supplementation at levels above 10 times the recommended dietary allowance has been documented for vitamin C (Rivers, 1987), vitamin E (Bendich and Machlin, 1988) and beta carotene (Bendich, 1988). There are very few other essential micronutrients which have been shown to beneficially affect immune responses without causing adverse effects at high dosages.

REFERENCES

Alexander, M., Newmark, H., and Miller, R. G., 1985, Oral betacarotenecinre se the number of OKT4+ cells in human blood, **Immunol. Lett.**, 9:221.

Alfin-Slater, R., Morris, R. S., 1963, Vitamin E and lipid metabolism, **Adv. Lipid Res.**, 1:183.

Alfin-Slater, R., and Morris, R. S., 1963, Vitamin E and lipid metabolism, **Adv. Lipid Res.**, 1:183.

Anderson, R., 1982, Ascorbic acid and immune functions: Mechanism of immunostimulation, *in*: **"Vitamin C Ascorbic Acid,"** J. N. Counsell, D. H. Hornig, eds., Applied Science, London.

Anderson, R., Lukey, P. T., 1987, A biological role for ascorbate in the selective neutralization of extracellular phagocyte-derived oxidants, **Ann. NY Acad. Sci.**, 498:229.

Anthony, L. E., Kurahara, C. G., and Taylor, K. B., 1979, Cell-mediated cytotoxicity and humoral immune response in ascorbic acid-deficient guinea pigs, **Am. J. Clin. Nutr.,** 32:1691.

Baehner, R. L., Boxer, L. A., Allen, J. M., and Davis, J., 1977, Autoxidation as a basis for altered function by polymorphonuclear leukocytes, **Blood,** 50:327.

Beisel, W. R., 1982, Single nutrients and immunity, **Amer. J. Clin. Nutr.,** 35:417.

Bendich, A., 1988a, Antioxidant vitamins and immune responses, pp. 125-147, *in:* **"Nutrition and Immunology,"** R. K. Chandra, ed., Alan R. Liss, Inc., New York.

Bendich, A., 1988b, A role for carotenoids in immune function, **Clin. Nutr.,** 7:113.

Bendich, A., D'Apolito, P., Gabriel, E., and Machlin, L. J., 1983, Modulation of the immune system function of guinea pigs by dietary vitamin E and C following exposure to oxygen, **Fed. Proc.,** 42:923.

Bendich, A., D'Apolito, P., Gabriel, E., and Machlin, L. J., 1984, Interaction of dietary vitamin C and vitamin E on guinea pig immune responses to mitogens, **J. Nutr.,** 114:1588.

Bendich, A., Gabriel, E., and Machlin, L.J., 1985, Role of dietary vitamin E on the natural killer cell lysis, XIII. International Congress on Nutrition, Brighton, UK.

Bendich, A., Gabriel, E., and Machlin, L. J., 1986a, Dietary vitamin E requirement for optimum immune responses in the rat, **J. Nutr.,** 116:675.

Bendich, A., Machlin, L. J., Scandurra, O., Burton, G. W., and Wayner, D. D. M., 1986b, The antioxidant role of vitamin C, **Advances in Free Radical Biology & Medicine,** 2:419.

Bendich, A., and Machlin, L. J., 1988, Safety of oral intake of vitamin E, **Am. J. Clin. Nutr.,** 48:612.

Bendich, A., and Shapiro, S. S., 1986, Effect of beta-carotene and canthaxanthin on the immune response of a rat, **J. Nutr.,** 116:2254.

Boxer, L. A., Oliver, J. M., Spielberg, S. P., Allen, J. M., and Schulman, J. D., 1979, Protection of granulocytes by vitamin E in glutathione synthetase deficiency, **New Eng. J. Med.,** 301:901.

Burns, J. J., 1959, Biosythesis of L-ascorbic acid: basic defect scurvy, **Am. J. Med.,** 26:740.

Burton, G. W. and Ingold, K. U., 1984, Beta-Carotene: an unusual type of lipid antioxidant, **Science,** 224:569.

Butterick, C. J., Baehner, R. L., Boxer, L. A., and Jersild, R. A., 1983, Vitamin E-A selective inhibitor of the NADPH oxidoreductase enzyme system in human granulocytes, **Am. J. Pathol.,** 112:287.

Castelli, A., Martorana, G. E., Meucci, E., and Bonetti, G., 1982, Vitamin C in normal human mononuclear and polymorphonuclear leukocytes, **Acta Vitaminol. Enzymol.,** 4:189.

Chavance, M., Brubacher, G., Herbeth, B., Vernhes, G., Mikstacki, T., Dete, F., Fournier, C., and Janot, C., 1984, Immunological and nutrtional status among the elderly, *in:* **"Lymphoid Cell Functions in Aging,"** A. L. de Weck, ed., Eurage, Rijswijk.

Chirico, G., Marconi, M., Colombo, A., Chiara, A., Rondini, G., and Ugazio, G., 1983, Deficiency of neutrophil phagocytosis in premature infants: effect of vitamin E supplementation, **Acta Paediatr. Scand.,** 72:521.

Chow, C. K., 1985, Vitamin E and blood, pp. 133-166, *in:* **"World Nutritional Determinants,"** G. H. Bourne, ed., Karger, Basil.

Colnago, G. L., Jensen, L. S., and Long, P. L., 1984, Effect of selenium and vitamin E on the development of immunity to coccidiosis in chickens, **Poultry Sci.,** 63:1136.

Cook, M. G., and McNamara, P., 1980, Effect of dietary vitamin E on dimethylnitrosamine-induced colonic tumors in mice, **Cancer Res.,** 40:1329.

Coquette, A., Vray, B., and Vanderpas, J., 1986, Role of vitamin E in the protection of the resident macrophage membrane against oxidative damage, **Arch. Int. Physiol. Biochem.,** 94:S29.

Corwin, L. M., and Shloss, J., 1980a, Influence of vitamin E on the mitogenic response of murine lymphoid cells, **J. Nutr.,** 110:916.

Corwin, L. M., and Shloss, J., 1980b, Role of antioxidants on the stimulation of the mitogenic response, **J. Nutr.,** 110:2497.

Corwin, L. M., and Gordon, R. K., 1982, Vitamin E and immune regulations, **Ann. NY Acad. Sci.,** 393:437.

Davey, F. R., and Dock, N. L., 1982, Effects of vitamin E on mixed lymphocyte cultures, **Ann. Nutr. Metab.,** 26:171.

Delafuente, J. C., Prendergast, J. M., and Modigh, A., 1986, Immunologic modulation by vitamin C in the elderly, **Int. J. Immunopharmoc.**, 8:205.

Deshago, R. D., Ewel, C., Londono, S., Metzger, Z., Hoffeld, J. T., and Openheim, J. J., 1981, Evidence for the involvment of monocyte-derived toxic oxygen metabolites in the lymphocyte dysfunction of Hodgkin's disease, **Clin. Exp. Immunol.**, 46:313.

Dillard, C. J., Kunert, K. J., and Tappel, A. L., 1982, Lipid peroxidation during chronic inflammation induced in rats by Freund's adjuvant: Effect of vitamin E as measured by expired pentane, **Res. Commun. Chem. Pathol. Pharmacol.**, 37:143.

Dobias, L., Lochman, I., Machalek, J., and Sram, R., 1986, Effects of ascorbic acid on humoral and other factors of immunity in coal-tar exposed workers, **J. Appl. Toxicol.**, 6:9.

Douglas, C. E., Chan, A. C., and Choy, P. C., 1986, Vitamin E inhibits platelet phospholipase A2, **Biochem. Biophys. Acta.**, 876:639.

Ellis, R. P., and Vorhies, M. W., 1986, Effect of supplemental dietary vitamin E on the serologic response of swine to an *Escherichia coli* bacterin, **J. Am. Vet. Med. Assn.**, 168:231.

Erickson, K. L., McNeill, C. J., Gershwin M.E, and Ossmann, J. B., 1980, Influence of dietary fat concentration and saturation on immune ontogeny in mice, **J. Nutr.**, 110:1555.

Erickson, K. L., Adams, D. A., and McNeill C.J., 1983, Dietary lipid modulation of immune responsiveness, **Lipids**, 18:468.

Eskew, M. L., Scolz, R. W., Reddy, C. C., Todhunter, D. A., and Zarkower, A., 1985, Effects of vitamin E and selenium deficiencies on rat immune function, **Immunol.**, 54:173.

Evans, R. M., Currie, L., and Campbell, A., 1982, The distribution of ascorbic acid between various cellular components of blood, in normal individuals, and its relation to the plasma concentration, **Brit J. Nutr.**, 47:473.

Fountain, M. W. and Schultz, R. D., 1982, Effects of enrichment of phosphatidylcholine liposomes with cholesterol or alpha-tocopherol on the response of lymphocytes to phytohemagglutinin, **Mol. Immunol.**, 19:59.

Fountain, M. W., Dees, C., Weete, J. D., and Schultz, R. D., 1982, Interactions of multilamellar phospholipid vesicles with biovine lymphocytes: Effects of alpha-tocopherol on lymphocyte blastogenesis, **Mol. Immunol.**, 19:1491.

Gallin, J. I., 1981, Abnormal phagocyte chemotaxis: Pathophysiology, clinical manifestations, and management of patients, **Rev. Infect. Dis.**, 3:1196.

Gebremichael, A., Levy, E. M., and Corwin, L. M., 1984, Adherent cell requirement for the effect of vitamin E on *in vitro* antibody synthesis, **J. Nutr.**, 114:1297.

Glatthaar, B. E., Hornig, D. H., and Moser, U., 1986, The role of ascorbic acid in carcinogenesis, *in*: **"Essential Nutrients in Carcinogenesis,"** L. A. Poirier, P.M. Newberne, M. W. Pariza, eds., Plenum Press, New York.

Goldschmidt, M. C., Masin, W. J., Brown, L. R., and Wyde, P. R., 1988, The effect of ascorbic acid deficiency on leukocyte phagocytosis and killing actinomyces viscosus, **Int. J. Vit. Nutr. Res.**, 58:326.

Gruner, S., Volk, H., Falck, P., and Von Baehr, R., 1986, The influence of phagocytic stimuli on the expression of HLA-DR antigens; role of reactive oxygen intermediates, **Eur. J. Immunol.**, 6:212.

Gurr, M. I., 1983, The role of lipids in the regulation of the immune system, **Prog. Lipid Res.**, 22:257.

Harris, R. E., Boxer, L. A., and Baehner, R. L., 1980, Consequences of vitamin E deficiency on the phagocyte and oxidative function of the rat polymorphonuclear leukocyte, **Blood**, 55:338.

Hatam, L. J. and Kayden, H. J., 1979, A high-performance liquid chromatographic method for the determination of tocopherol in plasma and cellular elements of the blood, **J. Lipid Res.**, 20:639.

Herberman, R. B., 1985, Natural killer (NK) cells: Characteristics and possible role in resitance against tumor growth, pp. 217-227, *in*:**"Immunity to Cancer,"** A. E. Reif, M. S. Mitchell, eds., Academic Press, Orlando, Florida.

Heinzerling, R. H., Tengerdy, R. P., Wick, L. L., and Lueker, D. C., 1974a, Vitamin E protects mice against Diplococcus pneumoniae type 1 infection, **Infect. Immun.**, 10:1292.

Heinzerling, R. H., Nockels, C. F., Quarles, C. L., Tengerdy, R. P., 1974b, Protection against chicks *E. coli* infection by dietary supplementation with vitamin E, **Proc. Soc. Ex. Biol. Med.**, 146:279.

Hemila, H., Roberts, P., and Wikstrom, M., 1985, Activated polymorphonuclear leucocytes consume vitamin C, **Febs Lett.**, 178:25.

Hill, H. R., Augustine, N. H., Rallison, M. L., and Santos, J. I., 1983, Defective monocyte chemotactic responses in diabetes mellitus, **J. Clin. Immun.**, 3:70.

Hoffeld, J. T., 1981, Agents which block membrane lipid peroxidation enhance mouse spleen cell immune activities *in vitro*: Relationship to enhancing activity of 2-mercaptoethanol, **Eur. J. Immunol.**, 11:371.

Hoffeld, J. T., 1983, Inhibition of lymphocyte proliferation and antibody production *in vitro* by silica, talc, bentonite or corynebacterium parvum: Involvement of peroxidative processes, **Eur. J. Immunol.**, 13:365.

Hruba, F., Novakova, V., and Ginter, E., 1982, The effect of chronic marginal vitamin C deficiencyon the alpha-tocopherol content of the organs and plasma of guinea pigs, **Experentia.**, 38:1454.

Hsu, C. K., 1977, Vitamin C and immune responses in rhesus monkey, **Fed. Proc.**, 36:1177.

Johnston, C. S., Cartee, G. D., and Haskell, B. E., 1985, Effect of ascorbic acid nutriture on protein-bound hydroxyproline in guinea pigs, **J. Nutr.**, 115:1089.

Kalden, J. R., and Guthy, E. A., 1972, Prolongedskin allograft survival in vitamin C-deficient guinea pigs, **Europ. Surg. Res.**, 4:114.

Kanazawa, K., Takeuchi, S., Hasegawa, R., Okada, M., Makiyama, I., Hirose, N., Toh, T., Cho, S. H., and Kobayashi, M., 1981, Influence of ascorbic acid deficiency on the level of non-protein SH compounds and vitamin E in blood and tissue of guinea pigs, **Nihon Univ. J. Med.**, 23:257.

Kennes, B., Dumont, I., Brochee, D., Hubert, C., and Neve, P., 1983, Effect of vitamin C supplements on cell-mediated immunity in old people, **Gerontology**, 29:305.

Kumar, M., and Axelrod, A. E., 1969, Circulating antibody information in scorbutic guinea pigs, **J. Nutr.**, 98:41.

Kurek, M. P., and Corwin, L. M.,1982, Vitamin E protection against tumor formation by transplanted murine sarcoma cells, **Nutr. Cancer**, 4:128.

Larson, H. J., and Tollersrud, S., 1981, Effect of dietary vitamin E and selenium on the phytohaemagglutinin response of pig lymphocytes, **Res. Vet. Sci.**, 31:301.

Lawrence, L. M., Mathias, M. M., Nockels, C. F., and Tengerdy, R. P., 1985, The effect of vitamin E on prostoglandin levels in the immune organs of chicks during the course of an E. coli infection, **Nutr. Res.**, 5:497.

Lehmann, L. J., and McGill, M., 1982, **J. Lipid Res.**, 23:299.

Leslie, C. A., and Dubey, 1982, D. P., Carotene and natural killer cell activity, **Fed. Proc.**, 41:331.

Levine, M., 1986, New concepts in the biology and biochemistry of ascorbic acid, **New Eng. J. Med.**, 314:892.

Likoff, R. O., Guptill, D. R., Lawrence, L. M., McKay, C. C., Mathias, M. M., Nockels, C. F., and Tengerdy, R. P., 1981, Vitamin E and aspirin depress prostaglandins in protection of chickens agains *Escherichia coli* infection, **Am. J. Clin. Nutr.**, 34:245.

Machlin, L. J., 1984, Vitamin E, *in*: "**Handbook of Vitamins,**" L. J. Machlin, ed., Marcel Dekker, Inc., New York.

Machlin, L. J. and Bendich, A., 1987, Free radical tissue damage: protective role of antioxidant nutrients, **FASEB J.**, 1:441.

Mbawuike, I. N., Rizzoni, W. E., Shloss, J., Kurek, M. P., and Corwin, L. M., 1982, Effect of K3T3 sarcomas on tissue concentrations of vitamin E, **Nutr. Cancer**, 2:140.

McMurray, D. N., 1984, Cell-mediated immunity in nutritional deficiency, **Prog. Food Nutr. Sci.**, 8:193.

Meade, C. J. and Mertin, J., 1978, Fatty acids and immunity, **Adv. Lipid Res.**, 16:127.

Meeker, H. C., Eskew, M. L., Scheuchenzuber, W., Scholz, R. W., and Zarkower, A., 1985, Antioxidant effects on cell-mediated immunity, **J. Leukocyte Biol.**, 38:451.

Mertin, J. and Hughes, D., 1975, Specific inhibitory action of polyunsaturated fatty acids on lymphocyte transformation induced by PHA and PPD, **Int. Archs Allergy Appl. Immunol.**, 48:203.

Mertin, J. and Hunt, R., 1976, Influence of polyunsaturated fatty acids on survival of skin allografts and tumor incidence in mice, **Proc. Nat. Acad. Sci. USA**, 73:928.

Meydani, S. N., Meydani, M., Verdon, C. P., Blumberg, J. B., Hayes, K. C., 1984, PGE2 control of vitamin E-enhanced immunity in old mice, **Fed. Proc.**, 43:478.

Meydani, S. N., Meydani, M., Verdon, C. P., Shapiro, A. A., Blumberg, J. B., and Hayes, K. C., 1986, Vitamin E supplementation supresses prostoglandin E Sythesis and enhances the immune system of aged mice, **Mech. Age and Devel.**, 34:191.

Meydani, S. N., Barklund, M. P., Liu, S., Meydani, M., Miller, R., Cannin, J., Morrow, F., Rocklin, R., and Blumberg, J., 1988, Effect of vitamin E supplementation on immune responsiveness of healthy elderly subjects, **Ann. NY Acad. Sci.**, in press.

Meydani, S. N., Blumberg, J. B., Yogeeswaran, G., and Meydany, M., 1989, Antioxidants and the aging immune system, *in*: "**Antioxidant Nutrients and Immune Functions**," A. Bendich, M. Phillips, R. Tengerdy, eds., Plenum Press, New York.

Morre, D. J., Crane, F. L., Sun, I. L., and Navas, P., 1987, The role of ascorbate in biomembrane energetics, **Ann. NY Acad. Sci.**, 498:153.

Moser, U. and Weber, F., 1983, Uptake of ascorbic acid by human granulocytes, **Int. J. Vit. Nutr. Res.**, 54:47.

Mueller, P. S., and Kies, M. W., 1962, Suppression of the tuberculin reaction in the scorbutic guinea pig, **Nature**, Lond., 195:183.

Mueller, P. S., Kies, M. W., Alvord, E. C. Jr., and Shaw, C. M., 1962, Prevention of experimental allergic encephalomyelitis (EAE) by vitamin C deprivation, **J. Exp. Med.**, 115:329.

Mulhern, S. A., Taylor, G. L., Magruder, L. E., and Vessey, A. R., 1985, Deficient levels of dietary selenium supress the antibody response in first and second generation mice, **Nutr. Res.**, 5:201.

Newberne, P. M., 1981, Dietary fat, immunological response, and cancer in rats, **Cancer Res.**, 41:3783.

Nockels, C. F., 1979, Protective effects of supplemental vitamin E against infection, **Fed. Proc.**, 38:2134.

Oberritter, H., Glatthaar, B., Moser, U., and Schmidt, K. H., 1986, Effect of functional stimulation on ascorbate content in phagocytes under physiological and pathological conditions, **Int. Archs. Allergy Appl. Immuno.**, 81:46.

Ohshima, H., and Bartsch, H., 1981, The influence of vitamin C on the *in vivo* formation of nitrosamines, pp.215-224, *in*: "**Vitamins, Nutrition, and Cancer**," J. N. Counsell and D. H. Hornig, eds., Applied Science, New Jersey.

Panush, R. S., Delafuente, J. C.,Katz, P., and Johnson, J., 1982, Modulation of certain immunologic responses by vitamin C III-Potentiation of *in-vitro* and *in-vivo* lymphocyte responses, **Int. J. Vit. Nutr. Res.**, 23:35.

Panush, R. S., and Delafuente, J. C., 1985, Vitamins and immunocompetence, 45:97, *in*: "**World Nutritional Determinants**," G. H. Bourne, ed., Karger, Basle.

Prasad, J. S., 1980, Effect of vitamin E supplementation on leukocye function, **Am. J. Clin. Nutr.**, 33:608.

RDA, 1980, Committee on Dietary Allowances, Food Nutrition Board, National Research Council: Recommended Daily Allowances, 9th ed., Washington, DC.

Reddanna, P., Rao, M. K., and Reddy, C. C., 1985, Inhibition of 5-lipoxygenase by vitamin E, **FEBS.**, 193:39.

Reddy, P. G., Morrill, J. L., Minocha, H. C., Morrill, M. B., Dayton, A. B., and Frey, R. A., 1985, Effects of supplemental vitamin E on the immune system of calves, **J. Dairy Sci.**, 69:164.

Rhodes, J., 1983, Human interferon action: reciprocal regulation by retinoic acid and beta carotene, **JNCI.**, 70:833.

Rhodes, J., Stokes, P., and Abrams, P., 1984, Human tumor-induced inhibition of interferon *in vitro*: reversal of inhibition of beta carotene (pro-vitamin A), **Cancer Immunol. Immunother.**, 16:189.

Roitt, I. M., 1984, "**Essential Immunology**," Blackwell Scientific Publications, London.

Schildknecht, E. G., and Squibb, R. L., 1979, The effect of vitamins A, E, and K on experimentally induced histomoniasis in turkeys, **Parasitology**, 78:19.

Scmidt, K. H., Steinhilber, D., Moser, U., and Roth, H.-J., 1988, L-ascorbic acid modulates 5-lipoxygenase activity in human polymorphonuclear leukocytes, **Int. Archs. Allergy Appl. Immun.**, 85:441.

Schwartz, J. L. and Shklar, G., 1989, Prevention and regression of hamster oral squamous cell carcinoma following administration of carotenoids, *in*: **"Antioxidant Nutrients and Immune Functions,"** A. Bendich, M. Phillips, R. Tengerdy, eds., Plenum Press, New York.

Schwartz, J., and Shklar, G.,1987, Regression of experimental hamster cancer by beta carotene and algae extracts, **J. Oral Maxillofac Surg.**, 45:510.

Schwartz, J., Suda, D., and Light, G., 1986, Beta carotene is associated with the regression of hamster buccal pouch caricinoma and the induction of tumor necrosis factor in macrophages, **Biochem. Biophy. Res. Comm.**, 136:1130.

Seifter, E., Rettura, G., and Leveson, S. M., 1981, Carotenoids and cell-mediated immune responses, *in*: **"Quality of Foods and Beverages: Recent Developments in Chemistry and Technology,"** G. Charalambous and G. Inglett, eds., Academic Press, New York.

Seifter, E., Rettura, G., Padawer, J., and Levenson, S. M., 1982, Moloney murine sarcoma virus tumors in CBA/J mice: Chemopreventive and chemotherapeutic actions of supplemental beta-carotene, **JNCI.**, 68:835.

Seifter, E., Rettura, G., Padawer, J., Stratford, F., Goodwin, P., and Levenson, S. M., 1983, Regression of C3HBA mouse tumor due to x-ray therapy combined with supplemental beta carotene or vitamin A, **JNCI.**, 71:409.

Seifter, E., Rettura, G., Padawer, J., Stratford, F., Weinzweig, J., Demetriou, A. A., and Levenson, S. M., 1984, Morbidity and mortality reduction by supplemental vitamin A or beta carotene in CBA mice given total-body gamma-radiation, **JNCI.**, 73:1167.

Shamberger, R. J., 1983, **"Biochemistry of Selenium,"** pp. 239-243, Plenum Press, New York.

Shapiro, A., Meydani, S. N., Macauley, J. B., Meydani, M., and Blumberg, J. B., 1984, Effect of vitamin E supplementation on plasma E level and lymphocyte proliferation in mice fed fish oil (FO), corn oil (CN), and coconut oil (CO) diets, **Fed. Proc.**, 44:1150.

Sharmanov, A. T., Aidarkhanov, B. B., and Kurmangaliev, S.M., 1986, Effect of vitamin-E on oxidative metabolism of macrophages, **Bull. Exper. Biol. Med.**, 101:810.

Sheffy, B. E., and Schultz, R. D., 1979, Influence on vitamin E and selenium on immune response mechanisms, **Fed. Proc.**, 38:2139.

Shilotri, P. G., 1977, Glycolytic, hexose monophosphate shunt and bactericidal activities of leukocytes in ascorbic acid deficient guinea pigs, **J. Nutr.**, 107:1507.

Shklar, G., Schwartz, J., Trickler, D. P., and Nuikian, K., 1987, Regression by vitamin E of experimental oral cancer, **JNCI.**, 78:987.

Shklar, G., and Schwartz, J., 1988, Tumor necrosis factor in experimental cancer regression with alphatocopherol, beta-carotene, canthaxanthin and algae extract, **Eur. J. Cancer Clin. Oncol.**, 24:839.

Siegel, B. V., 1974, Enhanced interferon responses to murine leukemia virus by ascorbic acid, **Infect. Immunol.**, 10:409.

Skopinska-Rozewska, E., Blaim, A., Wlodarska, B., Olszewski, M., and Galazka, B., 1987, The effect of vitamin E treatment on the incidence of okt lymphocytes in the peripheral blood of children with chronic respiratory tract infections, **Arch. Immun. Therap. Exper.**, 35:207.

Stephens, L. C., NcChesney, A. E., and Nockels, C. F., 1979, Improved recovery of vitamin E-treated lambs that have been experimentally infected with intratracheal chlamydia, **Br. Vet. J.**, 135:291.

Taccone-Gallucci, M., Giardini, O., Ausiello, C., Piazza, A., Bandino, D., Lubrano, R., Taggi, F., Evangelista, B., Monaco, P., Tabilio, M. R., Valeri, M., Citti, G., and Casciani, C.U., 1986, Vitamin E supplementation in hemodialysis patients: effects on peripheral blood mononeuclear cells lipid peroxidation and immune response, **Clin. Nephrology**, 25:81.

Teige, J., Tollersrud, S., Lund, A., and Larsen, H. J., 1982, Swine dysentery:The influence of dietary vitamin E and selenium on the xclinical and pathological effects of *Treponema hyodysenteriae* infection in pigs, **Res. Vet. Sci.**, 32:35.

Temple, N. J., and Basu, T. K., 1987, Protective effect of beta-carotene against colon tumors in mice, **JNCI.**, 78:1211.

Tengerdy, R., 1989, Feeding increased levels of vitamin E for immunity and disease resistance, *in*: **"Antioxidant Nutrients and Immune Functions,"** A. Bendich, M. Phillips, R. Tengerdy, eds., Plenum Press, New York.

Tengerdy, R. P., Nockels, C. F., 1975, Vitamin E or vitamin A protects chickens against *E. coli* infection, **Poultry Sci.**, 54:1292.

Tengerdy, R. P., Mathias, M. M., and Nockels, C. F., 1984, Effect of vitamin E on immunity and disease resistance, pp. 123-133, *in*: **"Vitamins, Nutrition, and Cancer,"** K. N. Prasad, ed., Karger, Basel.

Theron, A. and Anderson, R., 1985, Investigation of the protective effects of the antioxidants ascorbate, cysteine, and dapsone on the phagocyte-mediated oxidative inactivation of human alpha-1-protease inhibitor *in vitro*, **Am. Rev. Respir. Dis.**, 132:1049.

Theron, A., and Anderson, R., Investigation of the effects of oral administrationof ascorbate on the functional activity of serum alpha-1-protease inhibitor and oxidant release by blood phagocytes from cigarette smokers in a placebo-controlled, doubleblind, crossover trial, **Int. J. Nit. Nutr. Res.**, 58:218.

Thurman, G. B., and Goldstein, A. L., 1979, Suppression of immunological responsivity in guinea pigs by ascorbic acid, **Fed. Proc.**, 38:1173.

Till, G. O., Hatherill, J. R., Tourtellote, W. W., Lutz, M. J., and Ward, P. A., 1985, Lipid peroxidation and acute lung injury after thermal trauma to skin, **Am. J. Pathol.**, 119:376.

Tomita, Y., Himeno, K., Nomoto, K., Endo, H., and Hirohata, T., 1987a, Augmentation of tumor immunity against cyngeneic tumors in mice by beta carotene, **JNCI.**, 78:679.

Tomita, Y., Himeno, K., Nomoto, K., Endo, H., and Hirohata, T., 1987b, Augmentation of tumor immunity in mice by carotenoids [Abstract], Presented at Eighth International Symposium in Carotenoids, Boston.

Tvedten, H. W., Whitehair, C. K., 1973, Influence of the vitamins A and E on gnotbiotic and conventionally maintained rats exposed *Mycoplasma pulmonis*, **J. Am. Vet. Med. Assn.**, 163:605.

Vohra, K., Khan, A. J., Rosenfeld, W., Telang, V., and Evans, H. E., 1983, Correction of defective chemotaxis of neonatal neutrophils with ascorbic acid, **Pediatr. Res.**, 17:340.

Wattenberg, L. W., 1972, Inhibition of carcinogenic and toxic effects of polycylic hydrocarbonsby phenolic antioxidants and ethoxyquin, **JNCI.**, 48:1425.

Wayner, D. D. M., Burton, G. W., Ingold, K. U., and Locke, S., 1985, Quantative measurement of the total peroxyl radical-trapping antioxidant capability of human blood plasma by controlled peroxidation. The important contribution made by plasma proteins, **FEBS Lett.**, 187:33.

Weening, R. S., Schoorel, E. P., Roos, D., van Schaik, M. L., Voetman, A., Bot, A. A., Batenburg-Plenter, A. M., Willems, C., Zeijlemaker, W. P., Astaldi, A., 1981, Effect of ascorbate on abnormal neutrophil, platelet and lymphocyte function in a patient with Chediak-Higashi syndrome, **Blood.**, 57:856.

Zweiman, B., Besdine, R. W., and Hildreth, E. A., 1966, The effect of the scorbutic state on tuberculin hypersensitivity in the guinea pigs. II. *In-vitro* mitotic response of lymphocytes, **J. Immunol.**, 96:672.

ANTIOXIDANTS AND THE AGING IMMUNE RESPONSE

Simin Nikbin Meydani, Mohsen Meydani and Jeffrey B. Blumberg

USDA Human Nutrition Research Center on Aging
Tufts University, Boston, MA

INTRODUCTION

There is a growing recognition that nutrition influences immune function not only in young populations with severe malnutrition and a high incidence of infectious disease but also in groups with relatively mild or single nutrient deficiencies (Chandra and Chandra, 1986; James and Makinodan, 1984). Several comprehensive reviews in this area have been published recently (Stinnett, 1983; Gershwin et al., 1985; Fernandes, 1984; Beisel, 1982; Keusch, 1983). The elderly represent a large and expanding group with a significant number of individuals noted to possess poor nutritional status in the face of potentially increasing nutrient requirements and decreasing immunocompetence. Thus, as malnutrition impairs immunity, nutritional problems may contribute to declining immunity in old age and appropriate dietary intervention may improve immune responsiveness and reduce the burden of illness in the elderly. In an ecological milieu characterized by frequent illness and poor nutrition, the age-related decline in immune responsiveness may result in many of the chronic diseases associated with morbidity and mortality in the elderly. It also appears that conditions associated with chronic overnutrition, e.g., obesity, cardiovascular disease and adult-onset diabetes, significantly modulate immune function; the diets employed to treat these afflictions may have a far-ranging influence on host defense against infectious challenge. Recently, attention has been focused on the possibility of utilizing selective nutritional manipulations to regulate the aberrant response of diseases associated with immune disorders (Chandra, 1985; Corman, 1985).

One of the biological changes associated with aging is an increase in free radical formation with subsequent damage to cellular processes. Several studies have investigated the free radical theory of aging and the role of antioxidants, including vitamin E, on the life expectancy of rodents (see Blumberg and Meydani, 1986).

It has been suggested by Harman (1982) that vitamin E and other antioxidants may increase longevity by influencing the immune system and reducing age-related diseases. An immunological basis for many age-associated diseases such as amyloidosis, atherosclerosis, and cancer has been proposed by Walford, et al. (1981). Antioxidant supplementation has been shown to be protective against some of these diseases in animals, for example cancer (Horvath and IP, 1983) and amyloidosis (Meydani et al., 1986a). Furthermore, oxygen metabolites, especially H_2O_2 and oxidative products of arachidonic acid (AA), especially prostaglandin E_2 (PGE_2) produced by activated macrophages depress lymphocyte proliferation. Free radical formation associated with aging may be an underlying factor in the depressed immune response observed in aged rodents and improved antioxidant status might be beneficial in stimulating the immune response of the aged.

Immunological Changes Associated with Aging

Considerable evidence indicates that aging is associated with altered regulation of the immune system (Siskind, 1980). Well-documented, age-related functional changes have been defined for both humoral and cell-mediated responses (Kay 1979; Hallgren et al., 1973; Buckley et al., 1974). Although all four major cell types of the immune system, i.e., stem cells, macrophages, T-cells, and B-cells show age-related changes, the major alterations occur in the T-cells (Makinodan, 1981).

In vivo T-cell dependent, cell-mediated functions such as primary delayed hypersensitivity (Roberts-Thomson et al., 1974; Goodwin et al., 1982), graft vs. host reaction (Kay, 1979), and resistance to challenge with syngeneic and allogeneic tumors and parasites (Makinodan, 1981) are depressed with age. *In vitro*, the proliferative response of human and rodent lymphocytes to phytohemagglutinin (PHA) and Concanavalin A (Con A) as well as natural killer cell activity become depressed with age (Kay, 1979).

Cooperation between monocytes and lymphocytes is essential in antigen recognition, lymphocyte differentiation and eventual antibody production, and development of the effector state of cellular immunity or the delayed-hypersensitivity phase (Unanue, 1980). In addition to presenting antigen, macrophages synthesize interleukin-1 (IL-1) which induces the production of interleukin-2 (IL-2) by the activated T-cells. Macrophages have high levels of AA in their membrane phospholipids. Upon stimulation, mouse peritoneal macrophages release up to 50% of their AA content in the form of oxygenated metabolites; i.e., PG, hydroxyeicosatetraeonic acid (HETE), and leukotrienes (LT) (Humes et al., 1977; Bonney et al., 1985). These compounds, in addition to their effect on the biological activities of macrophages, suppress lymphocyte proliferation and lymphokine synthesis (Goodwin et al., 1974; Webb et al., 180; Gordon et al., 1976; Rola-Pleszezynski, 1985). Other oxidative metabolites of activated macrophages such as H_2O_2 have also been shown to suppress lymphocyte proliferation (Metzger et al., 1980; Zoschke and Messner, 1984).

Several groups have shown that antigen and mitogen stimulated IL-2 accumulation declines with age and contributes to the T-cell mediated defects observed with aging in rats, mice and humans (Thoman and Weigle, 1982; Miller and Stutman, 1981; Chang et al., 1982; Gillis et al., 1981). Miller (1983) using a limited dilution assay (LDA), showed a significant decrease with age in precursor frequencies of helper and cytolytic T-cells with no change in the magnitude of functional effect; i.e., IL-2 production and cytotoxicity produced per cell. However, with a LDA, unlike conventional assays, the effect of regulatory cells on IL-2 production is diluted.

Chang et al. (1982) showed that age-related changes in both murine macrophages and T-lymphocytes were responsible for decreased IL-2 production. Rosenberg et al. (1983) demonstrated that cell-cell interaction and cooperation via lymphokine and other regulatory molecules is impaired in aged mice and that increased macrophage numbers in aged rat spleen might have a suppressive effect. Chang et al. (1982) also showed that macrophages from old mice decreased IL-2 production by spleen non-adherent cells (NAC) from young mice and that culturing NAC from old mice with macrophages from young mice improved lymphocyte proliferation and IL-2 production. Bash (1983) showed that macrophages from young rats (in numbers up to 5% of NAC cells) enhanced lymphocyte proliferation and IL-2 formation while macrophages from old rats caused profound suppression (at 2.5% or above); the number of macrophages in spleen increased three-fold during aging. However, profound differences in the regulatory capacity of macrophages from young and old animals were observed when an equal number of macrophages were compared.

The suppressive effect of macrophages from aged mice has been attributed to either a decrease in IL-1 production (Chang, 1982) or an increase in suppressive factors. Increased PGE_2 production by macrophages from aged rats (Bash, 1983) and mice (Bartocci et al., 1982) has been reported. Furthermore, Rosenstein and Strauser (1980) were able to achieve substantial enhancement of aged spleen cell responsiveness *in vitro* and *in vivo* with indomethacin, a cyclooxygenase inhibitor. Bartocci et al. (1982) also showed that decreasing macrophage PGE_2 production with aspirin results in enhanced tumor rejection in aged mice. Splenocytes from aged mice synthesize more PGE_2 and accumulate less IL-2 in Con A-stimulated cultures than young mice (Meydani et al. 1986b).

Antioxidants and the Aging Immune Response

Vitamin E. Vitamin E is involved in maintenance of normal immune function. Tengerdy and Brown (1977) first reported that chickens given 100 mg/kg vitamin E had significantly increased generation of anti-sheep red blood cell (SRBC) plaque forming cells (PFC). Mice fed 60-100 mg/kg diets of vitamin E had significantly increased humoral immune responses as measured by PFC and antibody responses to SRBC and tetanus toxoid (Tengerdy, 1980). Vitamin E deficiency decreased the PFC response to SRBC in mice, an effect restored to normal by vitamin E but not by the antioxidant N-N-diphenyl-p-phenylenediamine (Tengerdy, 1980). Corwin and Shloss (1980a; 1980b) found that vitamin E and 2-mercaptoethanolamine were both mitogenic. Vitamin E supplementation in mice enhanced the proliferative response of lymphocytes to suboptimal doses of Con A. Pigs supplemented with vitamin E showed enhanced proliferation of the peripheral blastogenic response of lymphocytes to PHA (Larsen and Tollersrud, 1981). Vitamin E deficiency in dogs decreased the blastogenic response to Con A attributable to a serum factor that could be washed from the cell surface of depressed lymphocytes (Tanka et al., 1979). In mice, dietary vitamin E was shown to enhance helper T-cell activity. Bendich et al. (1983) reported that low splenic vitamin E levels in spontaneously hypertensive rats (SHR) were correlated with depressed splenic mitogen responses; tocopherol supplementation enhanced immune responsiveness in SHR and normotensive rats. Bendich et al. (1986) later demonstrated that 15 mg/kg diet/day of vitamin E was adequate to prevent myopathy in SHR rats but optimal lymphocyte proliferation to PHA and Con A was obtained only at vitamin E levels of 50 mg/kg diet/day. These studies indicate that the dietary tocopherol requirement for maintenance of optimal immune responsiveness may be higher than the levels recommended for normal growth and reproduction.

Garry et al. (1982), in assessing the nutritional status of a healthy elderly population, found that 25% consumed less than 50% of the RDA for tocopherol although most other reports do not indicate inadequate intake or status of vitamin E among the elderly (Leichter et al., 1978). Several studies have demonstrated an age-related increase in total serum tocopherol through the middle age (Chen et al., 1977; Wei Wo and Draper, 1975; Kelleher and Losowsky, 1978) followed by a decline after age 65 (Wei Wo and Draper, 1975; Barnes and Chen, 1981), which probably reflects similar changes in plasma lipid profiles (Horwitt et al., 1972). Although Vatassery et al. (1983) found that platelet vitamin E concentrations decline with age, Underwood et al. (1970) found no age-related change in liver tocopherol concentrations of people who died accidentally. Meydani et al. (1986) observed in rats that cerebellum and brain stem show selective decreases in tocopherol content with age. Lower serum tocopherol levels have been found in aged relative to young mice (Meydani et al., 1986b). An increase in average life span of short-lived autoimmune-prone NZB/NZW mice receiving vitamin E supplements was reported by Harman (1980).

These observations prompted us to evaluate the effect of vitamin E supplementation on the cell-mediated immune response of aged mice. We (Meydani et al., 1986b) found that 500 ppm dietary vitamin E supplementation of 24 month old C57BL/6j mice for 6 weeks significantly increased splenocyte proliferation to Con A and lipopolysachharide (LPS) but not to PHA relative to control animals fed 30 ppm of the vitamin (Figure 1). In addition, vitamin E supplementation significantly increased delayed cutaneous hypersensitivity (DCH) to 2, 4-dinitro-7-fluorobenzene (DNFB). This immunostimulatory effect of vitamin E was associated with an increased production of IL-2 and a decreased synthesis of PGE_2 (Table 1). No stimulatory effect of vitamin E was noted on natural killer cell (NK)-mediated cytotoxicity; however, when the mice were immunized with SRBC (a condition associated with increased oxidative stress) prior to assessment, the supplemented mice had a greater NK-mediated cytotoxicity (Meydani et al., 1988a) (Table 2).

Similar results were obtained by comparing the effect of an *in vitro* addition of tocopherol (4µg/ml in fetal bovine serum) on the mitogen-induced proliferative response of splenocytes from young and old mice fed corn oil or fish oil diets (Meydani and Blumberg, 1988b). While tocopherol alone was not mitogenic, it significantly enhanced the mitogenic response of splenocytes to PHA and Con A (Table 3). Vitamin E produced a greater response in young mice than in old mice on both diets. This effect might be due to a greater utilization and/or incorporation of tocopherol in the cells from young mice. Furthermore, a higher percent increase in mitogen-induced proliferation was observed in mice fed corn oil than those fed fish oil. This difference could be due to a higher tocopherol requirement associated with fish oil

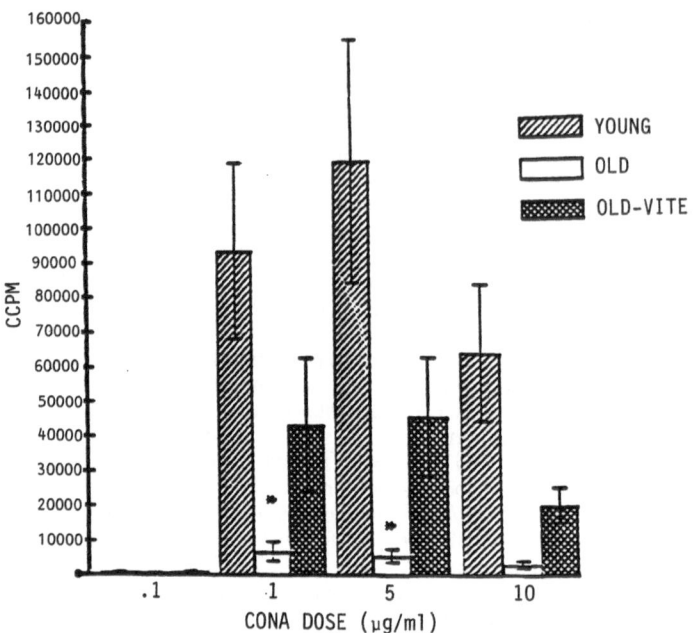

Figure 1. Effect of vitamin E supplementation on mitogenic response of mice splenocytes to Con A and PHA, reproduced from Meydani et al., 1986b with permission.

consumption (Meydani et al., 1987). These studies indicate that vitamin E supplementation improves the impaired immune response of aged mice. Further studies are required to determine the optimal level of vitamin E as well as the effect of longer term supplementation with vitamin E on the immune response of aged rodents.

Table 1. Effect of Vitamin E on immune responsiveness of 24 month old mice mean + SEM, N=9)a.

Immune Parameter	Dietary Vitamin E		
	30 ppm	500 ppm	p
DCH[b] (% change)	21.8 ± 6.1	45.1 ± 7.0	<0.05
IL-2[c] (units/ml)	12 ± 3	23 ± 6	<0.07
PGE$_2$[d] (mg/g wet weight)	3.20 ± 0.07	2.30 ± 0.10	<0.05

[a]Data adapted from Meydani et al., 1986b.

[b]Calculated by using below formula: E2-E1 / E1 x 100 where E1 is ear thickness prior toand E2 is ear thickness after DNFB challenge.

[c]Il-2 activity was measured in Con A-stimulated cultured spleen cells using a microassay described by Gillis et al. (1978).

[d]Ex-vivo synthesis of PGE$_2$ was measured by radioimmunoassay in spleen homogenates incubated for 10 minutes at 3°C.

Very few studies have examined the effect of vitamin E supplementation in humans. Goodwin and Garry (1983) in their survey of elderly subjects consuming megadoses of vitamin supplements did not see any correlation between vitamin E intake and tests of lymphocyte proliferation, and DCH. Interpretation of this survey is complicated by the fact that several vitamin supplements were used by each subject and the interaction between different nutrients present confounding variables. Harman and Miller (1986) supplemented 103 patients from a chronic care facility with 200 or 400 mg/day alpha-tocopheryl acetate but did not see any beneficial effect on antibody development against influenza virus vaccine. Unfortunately, data on the health status, medication use, antibody levels, and other relevant parameters were not reported. Chavance et al. (1985) conducted an epidemiological survey on the relationship between nutritional and immunological status in healthy French subjects over 60 years of age. They reported that plasma vitamin E levels were positively correlated with the number of positive DCH responses to diphtheria toxoid, candida and trichophyton. In men only, positive correlations were also observed between vitamin E levels and the number of positive DCH responses. Subjects with tocopherol levels greater than 135 mg/l were found to have higher helper-inducer/cytotoxic-suppressor ratios. Blood vitamin E concentrations were negatively correlated with the number of infectious disease episodes in three preceding years.

Table 2. Effect of vitamin E supplementation on NK-mediated cytotoxicity of old C57BL/6NNia mice fed corn oil (mean + SEM)a

Dietary vitamin E	SRBC injection	N	Target-effector ratio		
			1:100	1:50	1:25
mg/Kg			% specific lysis		
30	-	10	9.6 ± 1.2^b	5.6 ± 0.8^b	3.9 ± 0.6^b
30	+	4	4.2 ± 0.2	2.3 ± 0.1	1.3 ± 0.1
500	+	4	6.2 ± 1.9	5.1 ± 1.3	3.0 ± 0.9

[a]Reproduced from Meydani et al. (1988a) with permission.

[b]Significantly different at $p<0.05$ from mice fed 30 ppm vitamin E/Kg diet and injected with SRBC at $p<0.05$.

Table 3. Percent increase in mitogen-induced splenocyte proliferation by in-vitro addition of a-tocopherol (4mg/ml)a

Age	Diet	Percent Increase in Response	
		PHA	Con A
3 month	Corn oil	131	416
24 month	Corn oil	94	191
3 month	Fish oil	83	244
24 month	Fish oil	0	10

[a]Reproduced with permission from Meydani and Blumberg, 1988b. C57BL/6NNia mice were fed semi-synthetic diets containing 10% by weight of corn oil or 8.8% fish oil + 1.2% corn oil. Cells were incubated with tocopherol 4 hours prior to the addition of mitogens and then cultured for 72 hours. Percentages are calculated using cpm from cultures in the presence of a-tocopherol relative to those in the absence of a-tocopherol from the same animal.

We (Meydani et al., 1989) have recently completed a double blind, placebo-controlled clinical trial where healthy elderly subjects older than 60 years were supplemented with 800 IU/day of dl-α-tocopheryl acetate for a month. Each subject served as his/her own control. Plasma and white blood cell α-tocopherol content, DCH conducted with Multitest-CMI, mitogen-stimulated lymphocyte proliferation, as well as IL-1, IL-2, precursor frequencies for helper T-cells, PGE_2, and serum lipid peroxides were evaluated before and after treatment in addition to a comprehensive nutritional biochemistry profile. The data analyzed thus far indicate that in the vitamin E supplemented group: [1] the α-tocopherol content was significantly ($p.<0.001$) higher in plasma (3055 ± 1120 in supplemented vs 1029 ± 1028 µg/dl in placebo) and white blood cells (170 ± 52 in supplemented vs 80 ± 35 ng/10^7cells in placebo) than in the placebo group; [2] the cumulative diameter and number of positive antigens in DCH response were significantly ($p<0.05$) elevated; [3] the mitogenic response to optimal doses of the Con A but not to PHA were significantly ($p<0.05$) increased. Thus, this data suggests that vitamin E supplementation improves immune responsiveness in healthy elderly.

Vitamin C. Several studies have indicated that a low vitamin C intake (Hodkinson and Exton-Smith, 1976) or blood level (Wilson et al., 1972; Wilson et al., 1973) is associated with increased risk of death. However, in two randomized controlled trials, vitamin C supplementation of elderly people with low blood ascorbate levels did not decrease the mortality rate (Wilson et al., 1973; Burr et al., 1975). The failure of supplementation trials to show any beneficial effect may be due to the fact that irreversible damage had occurred as a result of a long-standing vitamin C deficiency and supplementation should have started earlier in life. On the other hand, low vitamin C status may have occurred as a consequence of poor health which ultimately caused death.

More direct evidence for a beneficial effect of vitamin C has been obtained by studying the effect of vitamin C supplementation on the immune response of elderly individuals. Kennes et al. (1983) examined the effect of intramuscular injections of vitamin C (500 mg/day for 1 month) on the proliferative response of lymphocytes to PHA and Con A and DCH response to tuberculin in 20 elderly subjects over the age of 70. A significant increase in ^{3}H-thymidine incorporation stimulated by PHA and Con A was observed after 30 days of supplementation. Vitamin C supplemented subjects also had an increase in the mean DCH induration diameter to tuberculin relative to placebo-treated subjects. As vitamin C status was not determined, it is not clear whether the observed improvement was due to correction of a vitamin C deficiency state or a direct immuno-stimulatory action of injected vitamin C. An immuno-stimulatory effect of vitamin C has been claimed in young people with presumably normal vitamin C levels (Anderson et al., 1980).

The mechanism of the immuno-stimulatory effect of vitamin C is not known. However, the serum level of lipid peroxides rises in healthy subjects with increasing age (Satoh, 1978; Svematsu et al., 1977, suggesting the immuno-stimulatory effect of vitamin C might be mediated through its antioxidant function. The level of certain antioxidant defenses such as vitamin C, selenium, glutathione and superoxide dismutase decrease with advancing age (Leibovitz and Siegel, 1980). Supplementation of elderly women with vitamin C or vitamin E for 12 months decreased serum peroxide levels by 13% and 26%, respectively (Wartanowicz et al., 1984). Vitamin C has been reported to increase *in vivo* generation of cyclic GMP (Atkinson et al., 1978), a signal for cell commitment into S phase (Katz et al., 1978).

Delafuente et al. (1986) studied a group of elderly patients over 65 years with chronic cardiovascular diseases receiving a variety of medications and examined the effect of *in vitro* and *in vivo* supplementation of vitamin C on lymphocyte proliferation and DCH to candida albicans and mumps skin test antigen. They found that, while *in vitro* addition of vitamin C to lymphocytes from elderly subjects increased their Con A-stimulated proliferation to levels comparable to those of young subjects, *in vivo* supplementation with 2 g/day vitamin C for 3 weeks did not significantly affect mitogenic responses or reverse anergy. These *in vivo* results are in contrast to those of Kennes et al. (1983) who studied healthy elderly subjects (receiving no medication) administered 500 mg/day of vitamin C intramuscularly and found improvement in the immunological parameters measured following supplementation. Unfortunately, plasma or white blood cell vitamin C levels were not measured in these studies.

Ziemlanski et al. (1986) found significantly increased serum IgG, IgM, and complement C-3 levels in 158 women over 78 years old receiving 400 mg ascorbic acid supplements. Goodwin and Garry (1983) found the healthy elderly subjects within the top 10% for plasma vitamin C concentration had significantly fewer anergic subjects responding to four different antigens and higher mean DCH scores. However, no difference in mitogenic response to PHA was observed between those with high and low vitamin C status.

In summary, lower plasma and leukocyte levels of vitamin C and age-related increases in serum lipid peroxides have been reported in the elderly. Compromised vitamin C status appears to contribute to the decreased immune responsiveness observed in the elderly although conflicting reports on the beneficial effects of high dose supplementation with this vitamin make unequivocal recommendations impossible.

Glutathione. Glutathione (GSH) is the most abundant low molecular weight thiol-containing compound in living cells. In its reduced form, GSH protects cells against various oxidants, free radicals, and cytotoxic agents (Sies and Wendell, 1978). Furthermore, it maintains a variety of cellular molecules in their functionally active form (Flohe and Gunzler, 1976). Recent studies indicate that GSH plays a role in lymphocyte activation. The depletion of GSH lowers mitogenic response and the addition of this tripeptide into culture medium reverses this effect (Noelle and Lawrence, 1981; Gougerst-Poidals et al., 1985). Decreased GSH levels as a function of age have been reported to occur in liver, kidney, heart, and blood of mice and have been suggested to be responsible for progression of the aging process (Hazelton and Lang, 1980; Abraham et al., 1978).

We (Furukawa et al., 1987) showed that 0.1% to 1.0% dietary GSH supplementation of a semipurified, nutritionally adequate diet significantly increased mitogenic response of aged mice to Con A and enhanced their *in vivo* T-cell mediated immune response as measured by DCH compared to that of aged mice fed control diet. Furthermore, they found that spleen and livers from aged mice had a lower GSH content than those from younger mice; GSH supplementation reversed this age-related decrease. Lacombe et al. (1985) showed that the level of intracellular - SH declines during mitogenic stimulation with Con A. However, Furukawa et al. (1987) found that old mice supplemented with GSH showed a smaller decline in intracellular GSH than control mice, suggesting that splenocytes from the aged mice supplemented with GSH maintain a more vigorous cellular GSH metabolism and are more responsive to mitogenic stimulation.

Methionine, a precursor of GSH biosynthesis has been reported not to have an immuno-stimulatory effect (Radix et al., 1983). An immuno-stimulatory effect in experimental animals has been reported for another sulfhydryl-containing compound, 2-mercaptoethanolamine (Heindrick et al., 1984); however, this antioxidant is not suitable for human use.

CONCLUSION

Despite the documented decline in immune responsiveness with age and the established changes in nutritional status in the elderly, little conclusive evidence is available about dietary interventions and immune function in the aged. Few studies with the exception of food restriction in animal models, have addressed the role of nutrient deficiencies or supplementation in the immune response of the aged.

Age-related increases in free radical reactions and associated lipid peroxidation events could contribute to the aging process via several mechanisms. Free radicals and the oxidative products of AA have been shown to have a suppressive effect on most cell-mediated immune functions. Dietary antioxidants may act to increase immune responsiveness by altering macrophage events mediated by cyclooxygenase, e.g., generation of PG or lipooxygenase, e.g., generation of HETE and LT or IL-1 production. The mechanism(s) of such action could be based upon antioxidant dampening of enzyme activity, quenching of lipid peroxidation, and/or altering fatty acid precursor pools of splenocyte phospholipids which dictate PG and LT synthesis.

The beneficial effect of three antioxidant nutrients, vitamin E, vitamin C, and GSH, lends support to the hypothesis that changes in lipid peroxidation and free radical formation with aging contributes to the observed functional alterations associated with the aging immune system. Furthermore, they indicate that improving the antioxidant status of the aged might be beneficial in retarding some of the age-related changes of the immune system and the subsequent development of age-associated diseases commonly found in the elderly. However, further long-term studies both in animals and humans are needed before scientifically supported recommendation can be provided for increased intake of dietary antioxidants.

REFERENCES

Abraham, C., Tal, Y., and Gershon, H., 1977, Reduced *in vitro* response to concanavalin A and lipopolysaccharide in senescent mice: a function of reduced number of responding cells, **Eur. J. Immunol.**, 7:301.

Atkinson, J., Kelly, J., Weiss, A., Wedner, H., and Parker, C., 1978, Enhanced intracellular CGMP concentrations and lectin-induced lymphocyte transformation, **J. Immunol.**, 121:2282.

Barnes, K. J., and Chen, L. H., 1981, Vitamin E status of the elderly in Central Kentucky, **J. Nutr. Elderly**, 1:41.

Bartocci, A., Maggi, F. M., Welker, A. I., and Veronese, F., 1982, Age-related immunosuppression:putatitve role of prostaglandins, *in*: "**Prostaglandins and Cancer**," T. J. Powles, R. S. Backman, K. V. Honn and P. Ramwell, eds., Alan R. Liss, New York.

Bash, J. A., 1983, Cellular immunosenescence in F344 rats; decline in responsiveness to phytohemagglutinin involves changes in both T cells and macrophages, **Mech. Aging Dev.**, 21:323.

Beisel, W. R., 1982, Single nutrients and immunity, **Am. J. Clin. Nutr.**, 35:417.

Bendich, A., Gabriel, E., and Machlin, L. J., 1983, Effect of dietary level of vitamin E on the immune system of the spontaneously hypertensive (SHR) and mormotensive Wistar Kyoto (WKY) rats, **J. Nutr.**, 113:1920.

Bendich, A., Gabriel, E., and Machlin, L. J., 1986, Dietary vitamin E requirement for optimum immune response in the rat, **J. Nutr.**, 116:675.

Blumberg, J. B., and Meydani, F. N., 1986, Role of dietary antioxidants in aging, *in*: "**Nutrition and Aging**," Vol. 5, H. Munro and M. Hutchinson, eds., Academic Press, New York.

Bonney, R. J., Opas, E. E., Humes, J. L., 1985, Lipoxygenase pathway of macrophages, **Fed. Proc.**, 44:2933.

Buckley, C. G., Buckly, E. G., and Dorsey, F. C., 1974, Longitudinal changes in serum immunoglobulin levels in older humans, **Fed. Proc.**, 33:2034.

Chandra, R. K., 1985, Nutritional regulation of immunity and infection: From epidemiology to phenomenology and clinical practice, **J. Pediatr. Gastroenterol. Nutr.**, 5:844.

Chandra, S., and Chandra, R. K., 1986, Nutrition, immune response and outcome, **Prog. Food Nutr. Sci.**, 10:1.

Chang, M. P., Makinodan, T., Peterson, W. J., Strehler, B. L., 1982, role of T cells and adherent cells in age-related decline in murine interleukin 2 production, **J. Immunol.**, 129:2426.

Chavance, M., Brubacher, G., Herberth, B., Vernes, G., Mistacki, T., Deti, F., Fournier, C., and Janot, C., 1985, Immunological and nutritional status among the elderly, *in*: "**Nutrition, Immunity and Illness in the Elderly**," R. K. Chandra, ed., Pergamon Press, New York.

Chen, C. H., Hsu, S. J., Huang, P. C., and Chen, J. S., 1977, Vitamin E status of Chinese population in Taiwan, **Am. J. Clin. Nutr.**, 30:728.

Corman, L. C., 1985, The relationship between nutrition, infection, and immunity, **Med. Clin. North Am.**, 69:519.

Corwin, L. M., and Shloss, J., 1980a, Influence of vitamin E on the mitogenic response of murine lymphoid cells, **J. Nutr.**, 110:916.

Corwin, L. M., and Shloss, J., 1980b, Role of antioxidants on the stimulation of the mitogenic response, **J. Nutr.**, 110:2397.

Delafuente, J. C., Dlesk, A., and Panush, R. S, 1981, Cellular immunity, *in*: "**Principles of Rheumatic Diseases**," R. S. Ranush, ed., John Wiley & Sons, New York.

Fernandes, G., 1984, Nutritional factors: Modulating effects on immune function and aging, **Pharm. Rev.**, 36:1235.

Flohe, L., and Gunzler, W. A., 1976, Glutathione-dependent oxido-reduction reactions, *in:* **"Glutathione: Metabolism and Function,"** I. M. Arias and W. B. Jakoby, eds., Raven Press, New York.

Furukawa, T., Meydani, S. N., and Blumberg, J. B., 1987, Reversal of age-associated decline in immune responsiveness, **Mech. Aging Dev.**, 3-8:107.

Garry, P. J., Goodwin, J. S., Hunt, W. C., Hooper, E. M., Leonard, A. G., 1982, Nutritional status in a healthy elderly population: dietary and supplemented intakes, **Am. J. Clin. Nutr.**, 36:319.

Gershwin, M. E., Beach, R. S., and Hurley, L. S., 1985, **Nutrition and Immunity,** Academic Press, Inc., Orlando, FL.

Gillis, S., Ferm, M. M., Ou, W., and Smith, K. A., 1978, T-Cell growth factor: Parameters of production and a quantitative microassay for activity. **J. Immunol.**, 120:2027.

Gillis, S., Kojak, R., Durante, and Weksler, M. E., 1981, Immunological studies of aging. Decreased production of and response to T cell growth factor by lymphocytes from aged humans. **J. Clin. Invest.**, 67:937.

Goodwin, J. S., and Garry, T. J., 1983, Relationship between megadose vitamin supplementation and immunological function in a healthy elderly population, **Clin. Exp. Immunol.**, 51:647.

Goodwin, J. S., Messner, R. P., and Peake, G. T., 1974, Prostaglandin suppression of mitogen stimulated leukocytes in culture, **J. Clin. Invest.**, 54:368.

Goodwin, J. S., Searles, R. P., and Tung, K.S.K., 1982, Immunological responses of a healthy elderly population, **Clin. Exp. Immunol.**, 48:403.

Gordon, D., Bray, M.,, and Morley, J., 1976, Control of lymphokine secretion by prostaglandins, **Nature**, 262:401.

Gougerst-Poidals, M. A., Fay, M., Roche, Y., Lacombe, P., and Marquetty, C., 1985, Immune oxidative injury in mice exposed to normabaric O_2: Effects of thiol compounds on the splenic cell sulfhydryl content and Con A proliferative response, **J. Immunol.**, 135:2045.

Hallgren, H. M., Buckley, C. E., Gilbertsten, V. A., and Yunis, E. J., 1973, Lymphocyte phytochemagglutinin responsiveness, immunoglobulins and autoantibodies in aging humans, **J. Immunol.**, 4:1101.

Harman, D., 1980, Free radical theory of aging: beneficial effect of antioxidants on the lifespan of male NZB mice: role of free radical reaction in the deterioration of the immune system with age and in the pathogenesis of systemic lupus erythematosus, **Age**, 3:64.

Harman, D., 1982, The free-radical theory of aging, *in:* **"Free Radicals in Biology,"** W. A. Pryor, ed., Academic Press, New York.

Harman, D., and Miller, R. W., 1986, Effect of vitamin E on the immune response to influenza virus vaccine and incidence of infectious disease in man, **Age**, 9:21.

Hazelton, G. A., and Lang, C. A., 1980, Glutathione contents of tissues in the aging mouse, **Biochem. J.**, 188:25.

Heindrick, M. L., Hendricks, L. C., and Cook, D. E., 1984, Effect of dietary 2-mercaptoethanol on the life span, immune system, tumor incidence and lipid peroxidation damage in spleen lymphocytes of aging $BC3F_1$ mice, **Mech. Aging Dev.**, 27:341.

Horowitt, M. K., Harvey, C. c., Dahm, C. J., Jr., and Searey, M. T., 1972, Relationship between tocopherol and serum lipid levels for determination of nutritional adequacy, **Ann. NE Acad. Sci.**, 203:223.

Horvarth, P. M., and Ip, C., 1983, Synergestic effect of vitamin E and selenium in the chemoprevention of mammary carcinogenesis in rats, **Cancer Res.**, 43:5335.

Hodkinson, H. M., and Exton-Smith, A. N., 1976, Factors predicting mortality in the elderly in the community, **Age and Aging**, 5:110.

Humes, J. L., Bonney, R. J., Pebes, L., et al., 1977, Macrophage synthesize and release prostaglandins in response to inflammatory response, **Nature**, 269:149.

James, S. J., and Makinodan, T., 1984, Nutritional intervention during immunologic aging: Past and present, *in:* **"Nutritional intervention in the Aging Process,"** H. J. Armbrecht, J. M. Prendergast, and R. M. Coe, eds., Springer-Verlag Press, New York.

Katz, S., Kierszenbaum, F., and Waksman, B., 1978, Mechanism of action of lymphocyte activating factor. III. Evidence that LAF acts on stimulated lymphocytes by raising cyclic GMP in Gl, **J. Immunol.**, 126:2386.

Kay, M.M.B., 1979, An overview of immune aging, **Mech. Aging Dev.**, 9:35.

Kelleher, J., and Losowsky, M. S., 1978, Vitamin E in the elderly, *in*: **"Tocopherol, Oxygen and Biomembranes,"** C. DeDuve, O. Hayaishi, eds., Elsevier/North Holland, Biomedical Press, Amsterdam.

Kennes, B., Dumont, I., Brohee, D., Hubert, C., and Neve, P., 1983, Effect of vitamin C supplementation on cell-mediated immunity in old people, **Gerontol.**, 29:305.

Keusch, G. T., Wilson, C. S., and Waksal, S. D., 1983, Nutrition, host defenses, and the lymphoid system, *in*: **"Advances in Host Defense Mechanisms,"** Vol. 2, J. I. Gallin and A. S. Fauci, eds., Raven Press, New York.

Lacombe, P., Kraus, L., Fay, M., and Pocidalo, J., 1985, Lymphocyte glutathione status in relation to their Con A proliferative response, **FEBS.**, 191:227.

Larsen, H. J., and Tollersrud, S., 1981, Effect of dietary vitamin E and selenium on the phytohaemagglutinin response of pig lymphocytes, **Am. J. Vet. Sci.**, 31:301.

Leibovitz, B. E., and Siegel, B. V., 1980, Aspects of free radical reactions in biological systems, **Aging J. Gerontol.**, 7;45.

Leichter, J., Angel, J. F., Lee, M., 1978, Nutritional status of a select group of free-living elderly people in Vancouver, **Can. Med. Assoc. J.**, 118:40.

Makinodan, T., 1981, Cellular basis of immunologic aging, *in*: **"Biological Mechanisms in Aging,"** USDA, NIH, R. T. Shimke. ed.

Metzger, Z., Hoffeld, J. T., and Oppenheim, J. J., 1980, Macrophage mediated suppression. I. Evidence of participation of both hydrogen peroxide and prostaglandins in suppression of murine lymphocyte proliferation, **J. Immunol.**, 124:983.

Meydani, M., McCauley, J., and Blumberg, J. B., 1986, Influence of dietary vitamin E, selenium and age on regional distribution of alpha-tocopherol in the rat brain, **Lipids**, 21:786.

Meydani, S. N., Cathcart, E. S., Hopkins, R. E., Meydani, M., Hayes, K. C., and Blumberg, J. B., 1986a, Antioxidants in experimental amyloidosis of young and old mice, *in*: **"Fourth Intl. Symposium of Amyloidosis,"** G. G. Glenner, E. P. Asserman, E. Benditt, E. Calkins, A. S. Cohen, D. Zucker-Franklin, eds., Plenum Press, New York.

Meydani, S. N., Meydani, M., Verdon, C. P., Shapiro, A. C., Blumberg, J. B., and Hayes, K. C., 1986b, Vitamin E supplementation suppresses prostaglandin E_2 synthesis and enhances the immune response in aged mice, **Mech. Aging Dev.**, 34:191.

Meydani, S. N., Shapiro, A. C., Meydani, M., McCauley, J., and Blumberg, J. B., 1987, Effect of age and dietary fat (fish oil, corn oil, and coconut oil) on tocopherol status of C57BL/6Nia mice, **Lipids**, 22:345.

Meydani, S. N., Yogeeswaran, G., Liu, S., Baskar, S., and Meydani, M., 1988a, Fish oil and tocopherol induced changes in natural killer cell mediated cytotoxicity and PGE_2 synthesis in young and old mice, **J. Nutr.**, 118:1245.

Meydani, S. N., and Blumberg, J. B., 1988b, Vitamin E and immune function in the elderly, *in*: **"Nutritional modulation of immune responses,"** S. Cunningham-Rundles, ed., Marcel Dekker Co. In Press.

Meydani, S. N., Barklund, M. P., Liu, S., Meydani, M., Miller, R., Cannon, J., Marrow, F., Rocklin, R., and Blumberg, J., 1989, Effect of vitamin E supplementation on immune responsiveness of healthy elderly subjects, **Ann. NY Acad. Sci.** In Press.

Miller, R. A., and Stutman, O., 1981, Decline in aging mice, of the anti-TNP cytotoxic response attributable to loss of Cyt-2-, IL-2-producing helper cell function, **Eur. J. Immunol.**, 11:751.

Miller, R. A., 1983, Age-associated decline in precursor frequency for different T-cell mediated reactions, with preservation of helper or cytotoxic effect per precursor cell, **J. Immunol.**, 132, 63.

Noelle, R. J., and Lawrence, D. A., 1981, Determination of glutathione in lymphocyte and possible association of redox state and proliferative capacity of lymphocytes, **Biochem. J.**, 198:571.

Radix, P. M., Walters, C. S., and Adkins, J. A., 1983, The influence of ethionine-supplemented soy protein diet on cell-mediated and humoral immunity, **J. Nutr.**, 113:159.

Rola-Pleszczynski, M., 1985, Immunoregulation by leukotrienes and other lipoxygenase metabolites, **Immunology Today**, 6:302.

Roberts-Thompson, I. C., Yvonschaiyud, V., and Whittingham, S., 1974, Aging, immune response and mortality, **Lancet II**:368.

Rosenberg, J. S., Gilman, S. C., and Feldman, J. D., 1983, Effect of aging on cell cooperation and lymphocyte responsiveness to cytokines, **J. Immunol.**, 130:1754.

Rosenstein, M. M., Strauser, H. R., 1980, Macrophage induced T-cell mitogen suppression with age, **J. Reticuloendoth. Soc.**, 27:159.

Satoh, K., 1978, Serum Lipid peroxides in cerebrovascular disorders determined by a new caloremetric methods, **Clin. Chim. Acta**, 90:37.

Sies, H., and Wendel, A., eds., 1978, **"Functions of Glutathione in Liver and Kidney,"** Springer-Verlag, New York.

Siskind, G. W., 1980, Immunological aspects of aging: an overview, *in*: **"Biological Mechanism in Aging,"** USDA, NIH, R. T. Schimke, ed.

Svematsu, T., Kamada, T., Abe, H., Kikudzi, S., and Yagi, K., 1977, Serum lipoperoxide level in patients suffering from liver disease, **Clin. Chim. Acta**, 79:267.

Tanka, J., Fuyiwara, H., and Torisu, M., 1979, Vitamin E and immune response: enhancement of helper T cell activity by dietary supplementation of vitamin E in mice, **Immunol.**, 38:727.

Tengerdy, P., and Brown, J. C., 1977, Effect of vitamin E and A on humoral immunity and phagocytosis in E coli infected chickens, **Poultry Sci.**, 56:957.

Tengerdy, R. P., 1980, Effect of vitamin E on immune responses, **Basic Clin. Nutr.**, 1:429.

Thoman, M. L., and Weigle, W. O., 1982, Cell-mediated immunity in aged mice: an underlying lesion in IL-21 synthesis, **J. Immunol.**, 128:2351.

Unanue, E. R., 1980, Cooperation between mononuclear phagocytes and lymphocytes in immunity, **New Engl. J. Med.**, 303:977.

Underwood, B. A., Sigel, H., Dolinski, M., and Weisell, R. C., 1970, Liver stores of alpha-tocopherol in a normal population dying suddenly and rapidly from unnatural causes in New York City, **Am. J. Clin. Nutr.**, 23:1314.

Vatassery, G. T., Johnson, G. J., and Krezowski, A. M., 1983, Changes in vitamin E concentrations in human plasma and platelets with age, **J. Am. Coll. Nutr.**, 4:369.

Walford, R. L., Gottesman, S.R.S., and Weindruch, R. H., 1981, Immunopathology of aging, **Ann. Rev. Gerontol. Geriatr.**, 2:3.

Wartanowicz, M., Panczenko-Kresowska, B., Ziemlanski, S., Kowalska, M., and Okolska, G., 1984, The effect of alpha-tocopherol and ascorbic acid on the serum lipid peroxide level in elderly people, **Am. J. Nutr. Metabol.**, 28:186.

Webb, D. R., Rogers, T. J., and Nowowiejski, I., 1980, Endogenous prostaglandin synthesis and the control of lymphocyte function, **Proc. NY Acad. Sci.**, 332:260.

Wei Wo, C. K., and Draper, H. H., 1975, Vitamin E status of Alaska Eskimos, **Am. J. Clin. Nutr.**, 28:808.

Ziemlanski, S., Wartanowicz, M., Panczenko-Kresowska, B., Kios, A., Kios, M., 1986, The effects of ascorbic acid and alpha-tocopherol supplementation on serum proteins and immunoglobulin concentrations in the elderly, **Nutr. Internatl.**, 2:1.

Zoschke, D. C., and Messner, R. P., 1984, Suppression of human lymphocyte mitogenesis mediated by phagocyte-released reactive oxygen species. Comparative activities in normal and in chronic granulomatous disease, **Clin. Immunol. Immunopath.**, 32:29.

ANTI-INFLAMMATORY SYSTEMS IN HUMAN MILK

Armond S. Goldman, Randall M. Goldblum, and Lars Å. Hanson

The University of Texas Medical Branch
Galveston, Texas
and
University of Goteborg, Goteborg, Sweden

INTRODUCTION

Mucosal surfaces of the alimentary tract and respiratory system are exposed to a host of deleterious agents, many of which are microorganisms and their products. The tissues of these organ-systems are defended by an array of resistance factors including the mucus barrier, lysozyme, lactoferrin, and antibodies principally of the secretory IgA isotype (Goldman et al., 1985; Udall, 1985). During early life, however, these immunologic mechanisms are poorly developed and the infant is therefore more vulnerable to the effects of microbial pathogens, foreign antigens, or inflammatory substances that are generated by the activation of elements of the next line of defense - the complement system, coagulation factors, certain isotypes of immunoglobulins (IgG, IgM, and IgE), local or elicited leukocytes, (mast cells, basophils, macrophages, neutrophils, NK cells, T cells) and a wide variety of mediators.

The susceptibility of human infants to those potential injurious agents is revealed in the absence of breast feeding. The incidence and severity of gastrointestinal and respiratory infections is much higher in non-breast fed infants, particularly in environments where there is considerable contact with intestinal pathogens (Woodbury, 1922; Grulee et al., 1935; Gordon et al., 1963; Cunningham, 1981; Mata et al., 1969; Mata et al., 1987; Mata and Urrutia, 1971; Duffy et al., 1986). Moreover, it is remarkable that breast fed infants exposed to pathogens such as enterotoxigenic *Escherichia coli, Shigella, Salmonella,* or rotavirus display little clinical evidence of inflammation in the aforementioned mucosal sites (Mata et al., 1969; Mata et al., 1987; Mata and Urrutia, 1971; Duffy et al., 1986). The most widely accepted explanation of this protection is that human milk supplies the very defense factors that the infant is slow to produce. A considerable body of information has accrued from descriptive studies to support that contention (Udall, 1985; Goldman and Goldblum, 1985; Goldman and Goldblum, 1988). Human milk is rich in lactoferrin (lactotransferrin), lysozyme, secretory IgA, anti-adherent oligosaccharides and glycoconjugates, anti-viral lipids, and other agents that either interfere with the multiplication or invasion of microorganisms or neutralize their toxins. Furthermore, because these defense agents are resistant to proteolysis (Goldman and Goldblum, 1985; Goldman and Goldblum, 1988; Lindh, 1975; Brines and Brock, 1983), they are well adapted to survive and operate in the hostile milieu of the alimentary tract. This paradigm may not, however, completely explain the paucity of inflammatory responses during the protective process. For that reason, a second model was proposed which predicted that human milk would be poor in the initiators or mediators of inflammation and rich in anti-inflammatory agents. An analysis of the scientific literature supported that proposition (Goldman et al., 1986).

PAUCITY OF INITIATORS AND MEDIATORS

The major biochemical pathways of the phlogistic systems are virtually absent in human milk. These include the complement components (Ballow et al., 1974), the coagulation system, the fibrinolytic factors, and kallikrien. A plasminogen activator has been described (Horie et al., 1982), but no substrate for this enzyme has been detected in human milk. Moreover, the major immunoglobulin isotypes (IgG, IgM, IgE) that aid in triggering inflammation are present in modest or very low concentrations, or in the case of IgE, are absent in human milk (Goldman and Goldblum, 1986; Underdown et al., 1976). Antigenic determinants of exogenous food antigens have been found in human milk (Kulangara, 1980; Stuart et al., 1984; Kilshaw and Cant, 1984; Harmatz et al., 1986), but so far intact foreign immunogens have not been isolated.

Human milk also contains many leukocytes, particularly in the first few months of lactation (Smith and Goldman, 1968; Ogra and Ogra, 1978; Goldman et al., 1982). Because most of the cells are neutrophils and macrophages, it would be anticipated that these leukocytes would injure the mucosal lining of the recipient after encountering suitable stimuli such as microorganisms or activated complement components. These cells have been found, however, to be refractory to certain bacterial ligands (N-formylmethionyl peptides) or the major chemoattractant - anaphylatoxin generated from the complement system, e.g., C5a (Thorpe et al., 1986). Other inflammatory cells, i.e., basophils, mast cells, eosinophils, platelets, natural killer cells and cytotoxic lymphocytes, are either not found or are poorly represented in human milk.

Finally, mediators of inflammation such as histamine, leukotrienes, and oxygen radicals have not been found in human milk. A peptide which is immunologically similar to the vasoactive intestinal peptide has been detected in human milk (Werner et al., 1985), but this agent may be more important in immunomodulation rather than in generating phlogistic responses.

ANTI-INFLAMMATORY FACTORS

General Aspects

The anti-inflammatory factors in human milk are not well represented in other mammalian milks that are currently used for infant nutrition. Many of the agents are also anti-microbial and are relatively resistant to digestion by the types of enzymes that pervade the gastrointestinal tract. Furthermore, the concentrations of those agents in human milk are highest during the initial phase of lactation, e.g., in colostrum.

Augmentation of Mucosal Barriers

Several different types of growth factors secreted into human milk may augment the growth of epithelial cells in the mucosa of the alimentary tract and respiratory system and thus increase the barrier effect of those lining structures. These growth factors include epidermal growth factor and mammary derived growth factor 1 and 2 (Werner et al., 1985; Klagsbrun, 1978; Carpenter, 1980; Moran et al., 1982; Kidwell et al., 1987). Indeed, certain animal model studies suggest that these growth factors may operate to increase the *in vivo* growth of intestinal epithelium (Widdowson et al., 1976; Hall and Widdowson, 1979; Heird and Hanson, 1977; Schwartz and Heird, 1981).

Specific secretory IgA antibodies, oligosaccharides and glycoconjugates by binding to microorganisms or their toxins are able to prevent their attachment to epithelial cells (Holmgren et al., 1976; Stoliar et al., 1976; Davis et al., 1982; Otnaess et al., 1983; Andersson et al, 1985; Holmgren et al., 1987). By excluding these foreign agents, sensitization and inflammation may be prevented. For example, the titers of serum IgG antibodies to cow's milk proteins are much lower in partially breast fed than completely cow's milk fed infants (Hanson et al., 1977) and the incidence of cow's milk sensitization appears to be lower during breast feeding.

Inhibition of Non-oxidative Systems

Human milk contains a number of agents that inhibit non-oxidant, inflammatory systems. Lactoferrin inhibits the complement system probably by modulating C3 convertase in the classical pathway of complement (Kijlstra and Jeurissen, 1982). Lysozyme and secretory IgA decrease the response of neutrophils to certain chemotactic agents in some *in vitro* models of leukocyte movement. Prostaglandins E2 and F2α inhibit neutrophil degranulation and lymphocyte activation (Weissmann et al., 1980), and pregnancy associated alpha 2-glycoprotein curbs lymphocyte blastogenesis (Horne et al., 1983). Enzymes that degrade histamine and leukotrienes (Heydicky, 1963; Blanc, 1981) and neutralizers of trypsin and chymotrypsin (α1-anti-trypsin and α1-anti-chymotrypsin) also are found in human milk (Lindberg et al., 1982) and should aid in preventing injury.

Inhibitors of Oxidants

Antioxidants are a significant part of the anti-inflammatory system in human milk (Table). Those agents inhibit oxidizing reactions by 1) decreasing the generation of superoxide by neutrophils, 2) sequestering available iron that might participate in Fenton or Fenton-like reactions that lead to the formation of hydroxyl radicals ($\cdot OH$) or other injurious oxygen intermediates, 3) degrading hydrogen peroxide, 4) inhibiting lipid peroxidation, and 5) scavenging toxic oxygen radicals. The lipid fraction of human milk has been reported to inhibit the generation of superoxide by human blood neutrophils (Pickering et al., 1980), but these lipid phase inhibitors have not been identified. Little free iron in human milk is available for Haberman-Weiss reactions. Lactoferrin,the most prominent whey protein in human milk, avidly binds Fe^{3+}. That protein in human milk is however largely unsaturated. This is of interest since iron saturated binding proteins support $\cdot OH$ formation and lipid peroxidation, whereas partially saturated ones inhibit those reactions (Baldwin et al., 1984; Winterbourn, 1983; Gutteridge et al., 1981; Aruoma and Halliwell, 1987). *In vitro* studies (Spik et al., 1982) reveal that the apoprotein is relatively resistant to proteolytic enzymes and balance studies suggest that some of this iron-binding protein from human milk persists throughout the alimentary tract of the recipient infant (Schanler et al., 1986). The fate of the agent in the recipient is not, however, well established.

When hydrogen peroxide is formed or lipid peroxidation occurs in the alimentary tract of the infant, human milk may aid in the catabolism of those compounds by providing catalase and glutathione peroxidase, respectively (Heydicky, 1963; Blanc, 1981). Furthermore, if toxic oxygen intermediates are produced at mucosal sites in the recipient infant, it is likely that they would be scavenged by the high concentrations of α-tocopherol (Jansson et al., 1981; Chapell et al., 1985; Ostrea et al., 1986), cysteine (Rassin, 1978), and ascorbic acid (Garza, 1982), and β-carotene (Chapell et al., 1985; Ostrea et al., 1986) found in colostrum and mature milk (Table 1).

Table 1. Antioxidants in Human Milk

Antioxidants	Functions
Lactoferrin	Binds Fe^{3+}. Prevents Fenton or Fenton-like reactions leading to OH formation.
Catalase	Degrades H_2O.
Glutathione peroxidase	Prevents lipid peroxidation.
β-carotene	Lipid antioxidant.
Cysteine	Scavenges radicals.
Ascorbate	Scavenges radicals. Regenerates reduced form of vitamin E.
Vitamin E α–tocopherol	Scavenges radicals. Immunostimulant.

Beta-carotene is apparently stored in the mammary gland during pregnancy and is rapidly secreted into the colostrum via milk fat globules and perhaps other mechanisms. Subsequently, the levels of that compound fall in transitional and more mature milk but the total amount secreted may not decline rapidly. The antioxidant activities of β-carotene have recently been reviewed (Burton and Ingold, 1984). At partial oxygen pressures that are less than 150 torr (e.g., as in normal atmospheric conditions), β-carotene is an excellent trapper of free radicals. Beta-carotene is not however a conventional chain-breaking antioxidant nor a peroxide-decomposing agent. It appears that β-carotene principally acts as a lipid antioxidant by complexing with the product of the reaction of carbon center radicals derived from lipid (often a polyunsaturated fatty acid) with oxygen. At high partial pressures of oxygen, β-carotene undergoes auto-oxidation. Therefore, one might predict that β-carotene in human milk might protect the mammary gland of the mother as well as the alimentary tract of the recipient against lipid peroxidation.

It is striking that the serum levels of this antioxidant as well as α-tocopherol are very low in umbilical cord blood apparently because of a specific barrier function of the placenta (Ostrea et al., 1986). After the infant receives the feedings of human milk for 4-6 days, the serum levels of these antioxidants reach adult levels (Ostrea et al., 1986). Human milk is necessary for the attainment of adult blood levels of vitamin E in early infancy (Ostrea et al., 1986). The serum levels of α–tocopherol in breast fed infants rise from a mean of 0.31 mg/dl at birth to about 0.9 mg/dl by the fourth day of life. Vitamin E in human milk may be of importance not only because of its direct antioxidant effects, but also because of its ability to potentially stimulate optimal immunologic responses (Tengerdy et al., 1981; Bendich et al., 1983; Bendich et al., 1984; Bendich et al., 1986).

The antioxidant effects of human colostrum upon the oxidant products of human neutrophils has been recently investigated in more detail (Buescher and McIlheran, 1988). Acellular colostrum prolonged the lag-time of hydrogen peroxide production, suppressed myeloperoxidase activity, and reduced cytochrome C activity. The antioxidant content of milk was estimated to be 918 ± 54 nmole/ml. The cytochrome C reducing activity of human milk was eliminated by ascorbate oxidase, whereas about 60% of the inhibition of the lag-time for hydrogen peroxide production was abolished. Ascorbate appeared to be a major antioxidant, but the system appears to be heterogeneous as previously suggested. The concomittant presence of ascorbic acid in human milk may have two purposes other than direct antixoxidant ones. The first may augment the immunostimulatory effects of vitamin E (Bendich et al., 1983; Bendich et al., 1984) and the second may be to react with the free radical form of vitamin E to regenerate its reduced form (Niki et al., 1982).

SUMMARY

Human milk is characterized not only by a complex host defense system that prevents the colonization and proliferation of common microbial pathogens that may pervade the alimentary tract and respiratory tract of the infant but also by a paucity of inflammatory agents and an array of anti-phlogistic factors. Clinical observations support the notion that the protection provided by human milk involves not only antimicrobial factors, but also anti-inflammatory agents. The major anti-inflammatory agents include enzymes that degrade mediators of inflammation, anti-proteases, lysozyme, lactoferrin, secretory IgA and a number of antioxidants including cysteine, ascorbate, α-tocopherol, and β-carotene. It is pertinent that most of these factors are either absent or poorly represented in cow's milk or other artificial feedings that substitute for breast feeding and that the attainment of adult serum levels of some of these antioxidants in early infancy is dependent upon breast feeding. It may be that the provision of these antioxidants may help to protect the recipient's developing immunologic system which is quite susceptible to oxidant damage. The absence of breast feeding will thus deprive the infant of valuable protection against common enteric-respiratory disorders and their inflammatory consequences.

It should be pointed out that the protective systems in human milk including the anti-inflammatory components may not be completely delineated, and that little is known of the *in vivo* fate of the factors and precisely how they protect the recipient. Those questions should form the basis of important research in the next decades.

Acknowledgements

This work was supported in part by a grant from the National Institute of Child Health and Human Development (5 R01 HD 21049-02) and the Swedish Medical Research Council (No. 215).

REFERENCES

Andersson, B., Porras, O., Hanson, L. Å., Svanborg-Eden, C., and Leffler, H., 1985, Nonantibody containing fractions of breast milk inhibit epithelial attachment of *Streptococcus pneumoniae* and *Haemophilus influenzae* , **Lancet,** I:643.

Aruoma, O. I., and Halliwell, B., 1987, Superoxide-dependent and ascorbate-dependent formation of hydroxyl radicals from hydrogen peroxide in the presence of iron. Are lactoferrin and transferrin promoters of hydroxyl-radical generation?, **Biochem. J.,** 241:273.

Baldwin, D. A., Jenny, E. R., and Aisen, P., 1984, The effect of human serum transferrin and milk lactoferrin on hydroxyl radical formation from superoxide and hydrogen peroxide, **J. Biol. Chem.,** 159:13391.

Ballow, M., Fang, F., Good, R. A., Day, N. K., 1974, Developmental aspects of complement components and C3 proactivator (properdin factor B) in human colostrum, **Clin. Exp. Immunol.,** 18:257.

Bendich, A., D'Apolito, P., Gabriel, E., and Machlin, L. J., 1983, Modulation of the immune system function of guinea pigs by dietary vitamins E and C following exposure to 100% O_2, **Fed. Proc.,** 42:923.

Bendich, A., D'Apolito, P., Gabriel, E., and Machlin, L. J., 1984, Interaction of dietary vitamin C and vitamin E on guinea pig immune responses to mitogens, **J. Nutr.,** 114:1588.

Bendich, A., Gabriel, E., and Machlin, L. J., 1986, Dietary vitamin E requirement for optimum immune responses in the rat, **J. Nutr.,** 116:675.

Blanc, B., 1981, Biochemical aspects of human milk - comparison with bovine milk, **World Rev. Nutr. Diet.,** 36:1.

Brines, R. D., and Brock, J. H., 1983, The effect of trypsin and chymotrypsin on the *in vitro* antimicrobial and iron binding properties of lactoferrin in human milk and bovine colostrum, **Biochem. Biophys.,** 759:229.

Buescher, E. S., and McIlheran, S. M., 1988, Further characterization of anti-oxidant components in human colostrum, **Pediatr. Res.,** 470A:1613 (abstract).

Burton, G. W., and Ingold, K. U., 1984, β-Carotene: An unusual type of lipid anti-oxidant, **Science,** 224:569.

Carpenter, G., 1980, Epidermal growth factor is a major growth-promoting agent in human milk, **Science,** 210:198.

Chapell, J. E., Francis, T., and Clandinin, M. T., 1985, Vitamin A and E content of human milk at early stages of lactation, **Early Hum. Dev.,** 11:157.

Cunningham, A. S., 1981, Breast-feeding and morbidity in industrialized countries: an update, *in*: **"Advances in International Maternal and Child Health,"** D. B. Jelliffe, E. F. P. Jelliffe, eds., Oxford University Press, New York.

Davis, C. P., Houston, C. W., Fader, R. C., Goldblum, R. M., Weaver, E. A., and Goldman, A. S., 1982, Immunoglobulin A and secretory immunglobulin A antibodies to purified type 1 *Klebsiella pneumoniae* pili in human colostrum, **Infect. Immun.,** 38:496.

Duffy, L. C., Riepenhoff-Talty, M., Byers, T. E., LaScolea, L. J., Zielezny, M. A., Dryja, D. M., and Ogra, P. L., 1986, Modulation of rotavirus enteritis during breast-feeding. Implication on alteration in the intestinal bacterial flora, **Am. J. Dis. Child.,** 140:1164.

Garza, C., Johnson, C. A., Harrist, R., and Nichols, B. L., 1982, Effects of methods of collection and storage of nutrients in human milk, **Early Hum. Dev.,** 6:295.

Goldman, A. S., Garza, C., Nichols, B. L., and Goldblum, R. M., 1982, Immunologic factors in human milk during the first year of lactation, **J. Pediatr.,** 100:563.

Goldman, A. S., and Goldblum, R. M., 1985, Protective properties of human milk, *in*: "Nutrition in Pediatrics - Basic Sciences and Clinical Application," W. A. Walker and J. B. Watkins, eds., Little, Brown and Co., Boston.

Goldman, A. S., and Goldblum, R. M., 1986, Immunoglobulins in human milk, *in*: "Natural Antimicrobial Systems," R. G. Board, ed., Bath University Press, Bath, U.K. and International Dairy Federation, Brussels.

Goldman, A. S., and Goldblum, R. M., 1989, Immunologic system in human milk: characteristics and effects, *in*: **"Textbook of Gastroenterology and Nutrition in Early Childhood,"** 2nd Edition, E. Lebenthal, ed., Raven Press, New York.

Goldman, A. S., Ham Pong, A. J., and Goldblum, R. M., 1985, Host defenses: Development and maternal contributions, *in*: **"Advances in Pediatrics,"** L. A. Barness, ed., Yearbook Medical Publ., Chicago.

Goldman, A. S., Thorpe, L. W., Goldblum, R. M., and Hanson, L. Å., 1986, Anti-inflammatory properties of human milk, **Acta Paediatr. Scand.**, 75:689.

Gordon, J. E., Chitkara, I. D., and Wyon, J. B., 1963, Weaning diarrhea, **Am. J. Med. Sci.**, 245:345.

Grulee, C. G., Sanford, H. N., and Schwartz, H., 1935, Breast and artificially-fed infants. A study of the age incidence in the morbidity and mortality in twenty thousand cases, **JAMA**, 104:1986.

Gutteridge, J. M. C., Patterson, S. K., Segal, A. W., and Halliwell, B., 1981, Inhibition of lipid peroxidation by the iron-binding protein lactoferrin, **Biochem. J.**, 199:259.

Hall, R. A., and Widdowson, E. M., 1979, Response of the organs of rabbits to feeding during the first days after birth, **Biol. Neonate.**, 35:131.

Hanson, L. Å., Ahlstedt, S., Carlsson, B., and Fallstrom, S. P., 1977, Secretory IgA antibodies against cow's milk proteins in human milk and their possible effect in mixed feedings, **Int. Arch. Allergy Appl. Immunol.**, 54:457.

Harmatz, P., Hanson, D. G., Brown, M., Kleinman, R. E., Walker, W. A., and Block, K. J., 1986, Transfer of maternal food proteins in milk, *in*: "Human Lactation. III. The Effects of Human Milk upon the Recipient Infant," A. S. Goldman, S. A. Atkinson, and L. Å. Hanson, eds., Plenum Press, New York and London.

Heird, W. C., and Hanson, I. H., 1977, Effect of colostrum on growth of intestinal mucosa, **Pediatr. Res.**, 11:406 (abstract).

Heydicky, G. V., 1963, Further investigations on the enzymes in human milk, **Pediatrics**, 31:1019.

Holmgren, J., Hanson, L. Å., Carlsson, B., Lindblad, B. S., and Rahimtoola, J., 1976, Neutralizing antibodies against *E. coli* and *V. cholerae* entero-toxin in human milk from a developing country, **Scand. J. Immunol.**, 5:867.

Holmgren, J., Svennerhold, A-M., Lindblad, M., Strecker, G., 1987, Inhibition of bacterial adhesion and toxin binding by glycoconjugate and oligosaccharide receptor analogues in human milk, *in*: **"Human Lactation. III: The Effects of Human Milk Upon the Recipient Infant,"** A. S. Goldman, S. A. Atkinson, and L. Å. Hanson, eds, Plenum Press, New York and London.

Horie, N., Okamoto, U., and Togawa, C., 1982, Studies on the plasminogen activating system in human milk. VI. Differences in properties between milk activator and glandular kallikrien, **Acta Haematol.**, 45:1099.

Horne, C. H. W., Armstrong, S. S., and Thomson, A. W., 1983, Detection of pregnancy associated α2-glycoprotein (α2-PAG), in IgA producing plasma cells and in body secretions, **Clin. Exp. Immunol.**, 51:631.

Jansson, L., Akesson, B., and Holmberg, I., 1981, Vitamin E and fatty acid composition of human milk, **Am. J. Clin. Nutr.**, 34:8.

Kidwell, W. R., Salomon, D. S., and Mohanam, S., 1987, Production of growth factors by normal human mammary cells in culture, *in*: **"Human Lactation. III. The Effects of Human Milk Upon the Recipient Infant,"** A. S. Goldman, S. A. Atkinson, and L. Å. Hanson, eds., Plenum Press, New York and London.

Kijlstra, A., and Jeruissen, S. H. M., 1982, Modulation of classical C3 convertase of complement by tear lactoferrin, **Immunology.**, 47:263.

Kilshaw, P. J., and Cant, A. J., 1984, The passage of maternal dietary proteins into human breast milk, **Int. Arch. Allergy Appl. Immunol.**, 75:8.

Klagsbrun, M., 1978, Human milk stimulates DNA synthesis and cellular proliferation in cultured fibroblasts, **Proc. Natl. Acad. Sci USA**, 75:5057.

74

Kolsto Otnaess, A-B., Laegreid, A., and Ertresvag, K., 1983, Inhibition of enterotoxin from *Escherichia coli* and *Vibrio cholerae* by gangliosides from human milk, **Infect. Immun.,** 40:563.

Kulangara, A. C., 1980, The demonstration of ingested wheat antigens in human breast milk, **IRCS Med. Sci.,** (Biochem) 8:(Part 1) 19.

Lindberg, T., Ohlsson, K., and Westrom, B., 1982, Protease inhibitors and their relations to protease activity in human milk, **Pediatr. Res.,** 16:479.

Lindh, E., 1975, Increased resistance of immunoglobulin dimers to proteolytic degradation after binding of secretory component, **J. Immunol.,** 114:284.

Mata, L. J., and Urrutia, J. J., 1971, Intestinal colonization of breast-fed children in a rural area of low socioeconomic level, **Ann. NY. Acad. Sci.,** 176:93.

Mata, L. J., Urrutia, J. J., Garcia, B., Fernandez, R., and Behar, M., 1969, Shigella infection in breast-fed Guatemalan Indian neonates, **Am. J. Dis. Child.,** 117:142.

Mata, L. J., Urrutia, J. J., and Gordon, J. E., Diarrheal disease in cohort of Guatemalan village children observed from birth to age two years, **Trop. Geogr. Med.,** 29:247.

Moran, R., Vaughn, R., and Orth, D. N., 1982, Epidermal growth factor (EGF) concentrations and daily production in breast milk during six weeks post delivery in mothers of premature infants, **Pediatr. Res.,** 16:172 (abstract).

Niki, E., Tsuchiya, J., Tanimura, R., and Kamiya, Y., 1982, Regeneration of vitamin E from alpha-chromanoxyl radical by glutathione and vitamin C, **Chem. Lett.,** 789.

Ogra, S. S., and Ogra, P. L., 1978, Immunologic aspects of human colostrum and milk. II. Characteristics of lymphocyte reactivity and distribution of E-rosette forming cells at different times after the onset of lactation, **J. Pediatr.,** 2:550.

Ostrea, E. A., Jr., Balun, J. E., Winkler, R., and Porter, T., 1986, Influence of breast-feeding on the restoration of the low serum concentration of vitamin E and β-carotene in the newborn infant, **Am. J. Obstet. Gynecol.,** 154:1014.

Pickering, L. K., Cleary, T. G., Kohl, S., and Getz, S., 1980, Polymorphonuclear leukocytes of human colostrum. I. Oxidative metabolism, **J. Infect. Dis.,** 142:685.

Rassin, D. K., Sturman, J. A., and Gaull, G. E., 1978, Taurine and other free amino acids in milk of man and other mammals, **Early Hum. Dev.,** 2:1.

Schanler, R. J., Goldblum, R. M., Garza, C., and Goldman, A. S., 1986, Enhanced fecal excretion of secreted immune factors in very low birth weight infants fed fortified human milk, **Pediatr. Res.,** 20:711.

Schwartz, S. M., and Heird, W. C., 1981, Further studies of colostrum-stimulated enteric mucosal growth, **Pediatr. Res.,** 15:546 (abstract).

Spik, G., Brunet, B., Mazurier-Dehaine, C., Fontaine, F., and Montreuil, J., 1982, Characterization and properties of the human and bovine lactotransferrin extracted from the faeces of newborn infants, **Acta Paediatr. Scand.,** 71:979.

Stuart, C. A., Twiselton, R., Nicholas, M. K., and Hide, D. W., 1984, Passage of cow's milk protein in breast milk, **Clin. Allergy.,** 14:533.

Smith, C. W., and Goldman, A. S., 1968, The cells of human colostrum. *In vitro* studies of morphology and function, **Pediatr. Res.,** 2:103.

Stoliar, O. A., Pelley, R. P., Kaniecki-Green, E., Klaus, M. H., and Carpenter, C. C., 1976, Secretory IgA against enterotoxins in breast milk, **Lancet,** I:1258.

Tengerdy, R. P., Mathias, M. M., and Nockels, C. F., 1981, Vitamin E, immunity and disease resistance, *in:* **"Diet and Resistance to Disease, "** M. Phillips and A. Baetz, eds., Plenum Press, New York.

Thorpe, L. W., Rudloff, H. E., and Goldman, A. S., 1986, The decreased response of human milk leukocytes to chemoattractant peptides, **Pediatr. Res.,** 20:373.

Udall, J. N., 1985, Immunologic aspects of gut function, *in:* **"Nutrition in Pediatrics. Basic Science and Clinical Application,"** W. A. Walker and J. B. Watkins, eds., Little, Brown, and Company, Boston/Toronto.

Underdown, B. J., Knight, A., and Papsin, F. R., 1976, The relative paucity of IgE in human milk, **J. Immunol.,** 116:1435.

Weissmann, G., Smolen, J. E., and Korchak, H., 1980, Prostaglandins and inflammation: Receptor/cyclase coupling as an explanation of why PGEs and PGI_2 inhibit function of inflammatory cells, *in:* **"Advances in Prostaglandin and Thomboxane Research,"** B. Samuelsson, P. W. Ramwell, and R. Paoletti, eds., Raven Press, New York.

Werner, H., Koch, Y., Fridkin, M., Fahrenkrug, J., and Gozes, I., 1985, High levels of vasoactive intestinal peptide in human milk, **Biochem. Biophys. Res. Commun.**, 133:228.

Widdowson, E. M., Colombo, V. E., and Artavanis, C. A., 1976, Changes in the organs of pigs in response to feeding for the first 24 hours after birth. II. The digestive tract, **Biol. Neonate**, 28:272.

Winterbourn, C. C., 1983, Lactoferrin-catalysed hydroxyl radical production. Additional requirement for a chelating agent, **Biochem. J.**, 210:15.

Woodbury, R. M., 1922, The relation between breast and artificial feeding and infant mortality, **Am. J. Hyg.**, 2:668.

THE ADMINISTRATION OF BETA CAROTENE TO PREVENT AND REGRESS ORAL CARCINOMA IN THE HAMSTER CHEEK POUCH AND THE ASSOCIATED ENHANCEMENT OF THE IMMUNE RESPONSE

Joel L. Schwartz, Gerald Shklar, Evelyn Flynn, and Diane Trickler

Harvard School of Dental Medicine
Department of Oral Pathology and Oral Medicine
188 Longwood Ave.
Boston, MA 02115

SUMMARY

In the past four years this laboratory has utilized the hamster cheek pouch tumor model to investigate the anticancer activities of antioxidants, such as beta carotene. These molecules, which have exhibited no evidence of toxicity, have been administered systemically (oral ingestion), and locally to the tumor site in the hamster cheek pouch. The results have been either the inhibition of tumor growth, or the regression of tumor. Adjacent to the degenerating tumors a dense inflammatory infiltrate was observed. Specifically, the cytokines, tumor necrosis factor alpha, and beta, have been immunohistochemically localized to the site of regressed oral carcinoma. Recently, liposomes composed of phosphaditylcholine, phosphaditylserine, and phosphodityelanolamine were combined with beta carotene and injected locally to oral squamous cell carcinoma of the hamster. The results indicated that tumor cells accumulated the liposomes and were lysed while normal mucosal cells did not demonstrate this effect. Therefore antioxidants such as beta carotene can be localized to a tumor site, without a toxic response. Future studies on the anticancer activity of the antioxidants need to focus on the cellular and molecular changes produced in the immune effectors and in the mucosal cells following administration of the antioxidants.

INTRODUCTION

Previous findings *in vivo* and *in vitro* have shown that the oral administration of the carotenoid, beta carotene, can inhibit the growth of mouse mammary carcinoma (Seifter et al., 1984), hamster salivary carcinoma (Alam et al., 1984) and C3HBA inoculated adenocarcinoma in the mouse (Rettura et al., 1982).

In terms of photoreactive studies, Mathews-Roth has carried out a series of studies with the oral administration of beta carotene in mice, demonstrating that a diet high in beta carotene resulted in the accumulation of beta carotene in the skin of the animal (Mathews-Roth, 1982). In another study she demonstrated that the oral administration of beta carotene was associated with an increase in the latent period of tumor development and growth (Mathews-Roth and Krinsky, 1984). These results have been confirmed recently by Nishino et al., (1988) who has shown that alpha and beta carotene suppressed skin tumor promotion by 12-tetradecanoyl-phorbol-13-acetate (TPA) in DMBA initiated mice. Beta carotene and canthaxanthin, a carotenoid that does not convert to vitamin A have been shown to inhibit the transformation of the 10T1/2 fibroblast cell line. The inhibition, supposedly by antioxidant properties of the carotenoids was suggested to be via a mechanism different than retinoid inhibition (Pung et al., 1988). This laboratory has produced data in the hamster, demonstrating inhibited promotion of hamster tumor by benzoyl peroxide (40.0%) and, to some extent, initiation by DMBA (Suda and Schwartz, 1986).

In terms of immune responsiveness, the oral administration of beta carotene has been associated with the induction and the activation of cytotoxic T lymphocytes (Tomita et al., 1987). These lymphocytes could be adoptively transferred to inhibit a meth -A induced fibrosarcoma. (Thy-1 positive, Lyt-1 negative, Lyt-2 positive). The mitogenic activity of the carotenoids, beta carotene, and canthaxanthin has also been documented by Bendich and Shapiro (1986). Further studies by Rhodes, have shown that, while human carcinoma inhibited macrophage cytotoxicity and gamma interferon production, this inhibition was reversed following incubation with beta carotene (Rhodes, 1983; Rhodes et al., 1981). Recent studies with various human cancer populations have indicated that antioxidants such as beta carotene if found in high levels in the serum, could reduce the risk of carcinoma development of the lung (Menkes et al., 1986), prostate (Ohno et al., 1988), bladder (Wald et al., 1988), bladder (Wald, 1982), and oral mucosa (Stich et al., 1986; Stich and Dunn, 1986; and Stich et al., 1988). The purpose of the studies conducted in this laboratory was to provide evidence of the relationship of antioxidants such as beta carotene and canthaxanthin to their capacity to alter the growth of oral squamous cell carcinoma of the hamster cheek pouch tumor model.

MATERIALS AND METHODS

Animals

Random bred male Golden Syrian hamsters (Mesocricetus auratus, Lakeville strain LVG - Charles River Breeding Laboratories, Wilmington, MA) 4-5 weeks of age, certified free of parasites, were fed a rat chow supplemented with the required vitamin E and A concentrations, 22% of their diet protein, (Ralston Purina rat chow, #5012) and given water ad libitum. The hamsters received 12 hours of light and 12 hours of dark. Each hamster normally eats 5-7 gm per day, drinks 10 ml water, and weighs 90-120 gms at the start of the experiment. A total of 917 hamsters were used in these studies described below.

Tumor Model

The complete carcinogen 7,12 dimethylbenz(α)-anthracene (DMBA) was applied to the hamster buccal mucosa three times a week using a #4 sable brush. The concentration of the carcinogen varied depending on the carcinogenesis process the investigator wished to produce. In the initiation-promotion studies and in the prevention studies, a 0.25% concentration of DMBA in heavy mineral oil was applied 3 times per week respectively for 10 or 21 weeks. In the inhibition studies the concentration varied from 0.25% to 0.1% applied for 22 weeks. To establish tumors the concentration that was used was 0.5% applied for 14 weeks. The concentration of the carcinogen has been determined to be 0.4 mg per application using ^{14}C radiolabelled DMBA (New England Nuclear, Boston, MA). In all instances the experimental antioxidant was administered on days alternate to the carcinogen application. The application of carcinogen results in the transformation of the hamster buccal mucosal cells. This process can be visualized by the initial development of hyperkeratosis, then the formation of the premalignant lesion, leukoplakia, in some cases associated with erthroplasia, papillary carcinoma and then invasive cancer development (Shklar, 1972). The process of carcinogenesis or regression can be monitored by inverting the hamster pouch following the induction of light anesthesia with carbon dioxide, and recording the lesions or tumor present in the pouch.

Carcinoma Cell Line

The cloned cell line HCPC-1 was established from an epidermoid carcinoma of the Syrian hamster. The carcinoma was induced by the application of the carcinogen DMBA. The cell line has been maintained through over 100 passages for five years and repeatedly cloned over that period. Electron microscopic analysis of the cell line revealed tonofilaments and desmosomes, thereby confirming its epithelial origin. Presence of keratin in the HCPC-1 was demonstrated by both histochemical and indirect immunofluorescent studies. The population-doubling time of HCPC-1 was estimated from the exponential growth phase of the growth curve established for the cell line, which was found to be 12 hours. The plating efficiency of the cell line was estimated at 35%. Antigenicity was determined to be low; a successful transplant into the hamster cheek pouch or into the peritoneum of nonimmunosuppressed animals required

3.6 X 10^7 cells to produce an inflammatory infiltrate sufficient for rejection of the established tumor. Successful transplants were achieved with 10^6 cells injected into hamster pouch or peritoneum of nonimmuno-suppressed hamsters (Odukoya et al., 1983).

Inhibition Studies

Two experiments were performed to determine the inhibition capability of beta carotene on the process of carcinogenesis (Suda and Schwartz, 1986).

In the first experiment, forty adult male Syrian hamsters were divided into four equal groups with 10 animals in each group. The groups were as follows:

Group 1. DMBA, Group 2. DMBA + Beta carotene, Group 3. Beta carotene, and Group 4. No treatment

Group 1 animals were painted on the left buccal pouch thrice weekly with a No. 4 sable brush. Group 2 animals were painted thrice weekly with DMBA as in group 1. On alternate days the left buccal pouch was painted with a 2.5% solution of β-carotene (Sigma St. Louis, MO.) dissolved in mineral oil. Each painting placed approximately 0.62 mg. of β-carotene on the mucosal surface of the pouch. The other groups were treated as stated above. The animals were not restrained or anesthetized during the application of the carcinogen or the delivery of β-carotene.

The second experiment involved the effect of beta carotene on early and late phases of carcinogenesis (initiation and promotion).

Inititiation and Promotion Model

The model for initiation and promotion in the hamster buccal pouch was developed by Odukoya and Shklar (Odukoya and Shklar, 1984), based upon the experimental system developed for the skin by Slaga and associates (Slaga et al., 1981). Initiation was induced using a 0.25% DMBA solution applied for ten weeks. This was by a six-week period of rest, followed by another nine weeks of treatment with 40.0% benzoyl peroxide in acetone.

Eighty young adult male Syrian hamsters similar to those in experiment 1 were divided into eight equal groups of 10 animals.

Group 1. DMBA (10 wks)-rest (6 wks)-40.0% benzoyl peroxide (9 wks).
Group 2. DMBA (10 wks)+190 ng/0.1 ml, β-carotene, in mineral oil painted on days alternate to the carcinogen three times per week. Rest (6 wks) was followed by 40.0% benzoyl peroxide (9 wks).
Group 3. DMBA (10 wks)-rest (6 wks)+β-carotene applied as above - 40.0% benzoyl peroxide (9 wks).
Group 4. DMBA (10 wks)-rest (6 wks)-40.0% benzoyl peroxide (9 wks) + β-carotene.
Group 5. DMBA (10 wks)+mineral oil-rest (6 wks)-40.0% benzoyl peroxide (9 wks).
Group 6. DMBA (10 wks)-rest (6 wks) + mineral oil -40.0% benzoyl peroxide (9 wks).
Group 7. DMBA (10 wks)-rest (6 wks)-40.0% benzoyl peroxide (9 wks) + mineral oil.
Group 8. Untreated.

Prevention Study with Carotenoids

140 male Syrian hamsters were painted with 0.1% DMBA in heavy mineral oil, three times per week for six months. Oral squamous cell carcinoma was induced between 24-28 weeks. The animals were divided into seven equal groups of 20 animals. Carotenoids were administered by mouth (plastic 1 cc syringe) three times per week on days alternate to the carcinogen. Each feeding of the carotenoids was 1.4 mg/kg. dissolved in 0.4 ml. mineral oil. The groups were as follows:

Group 1. DMBA
Group 2. DMBA + mineral oil
Group 3. DMBA + carotenoid mixture
Group 4. DMBA + beta carotene

Group 5. DMBA + canthaxanthin (CTX)
Group 6. normal
Group 7. normal + carotenoid mixture

Gamma Glutamyl Transpeptidase

Levels of activity for the membrane bound enzyme gamma glutamyl transpeptidase (GGT) were studied in an inhibition study using DMBA tumor induction and beta carotene. Forty hamsters were divided into four separate groups (Suda et al., 1987). The groups were as follows:

Group 1. DMBA, 0.25% painted thrice weekly for 22 weeks.
Group 2. DMBA + beta carotene, painted as above with DMBA, and on alternate days painted with an approximate 0.25% solution of beta carotene dissolved in mineral oil. (250 ug/0.25 ml., 520.8 ug./kg.)
Group 3. Beta carotene, administered in the amount stated above.
Group 4. Untreated.

GGT Assay

The inverted hamster pouches were removed and fixed in 1.0% glacial acetic acid. The whole mount method was prepared according to the technique used by Solt (1981). After remaining in an ice cold aqueous solution of 1.0% acetic acid with constant stirring for 1 hour, the pouch tissues were rinsed and placed in a shallow pan of distilled water, and the epithelium detached from the underlying connective tissue with a wooden tongue depressor. The epithelial fragments were then placed, basal layer up on glass slides, air dried and secured to the glass surface with glycerin gelatin. Localization of GGT within the whole mount was achieved using the histochemical method of Rutenberg et al. (1969). In order to quantitate the number of GGT-stained lesions, the outline of each whole mount preparation was traced onto graph paper with sq. mm. divisions. The number of GGT-stained lesions was determined microscopically and expressed as a function of epithelial surface area (Solt and Shklar, 1982).

Regression Studies

Four separate studies were conducted in which established DMBA oral carcinomas were regressed or inhibited in their growth following the administration of carotenoids (Schwartz, 1988; Schwartz and Shklar, 1988) (Study one). Fifty randomly bred male Syrian hamsters developed oral carcinoma following the application of 0.25% of DMBA, 3 times a week for 22 weeks. In addition, fifty hamsters were injected with 2.5×10^6 cells/0.5 cc injected subcutaneously (#26 gauge needle) into the buccal pouches of non-immunosuppressed 70-80 week-old hamsters, tumors were visible after four days. Beta carotene was administered, 190 ug/0.1 ml., by injection two times per week for four weeks. Beta carotene was also applied in a mineral oil suspension at an approximate concentration of 250 ug/ml on days alternate to the carcinogen, three times a week. The controls for the beta carotene activity were 13-cis retinoic acid, 190 ug/0.1 ml. (Hoffman-La Roche, Nutley, NJ) and canthaxanthin (Hoffman-La Roche), 190 ug./0.1 ml. The following groups were used to determine the effect of topical application of beta carotene on DMBA induced tumors in the hamster pouch.

Group 1. DMBA
Group 2. DMBA + beta carotene
Group 3. Beta carotene
Group 4. DMBA + mineral oil
Group 5. Mineral oil

The groups for the study of the injection of beta carotene locally to the HCPC-1 tumor were as follows:

Group 1. HCPC-1
Group 2. HCPC-1 + beta carotene
Group 3. HCPC-1 + Medium

Group 4. Medium
Group 5. Saline (PBS)

In three other studies beta carotene, canthaxanthin, and 13-cis retinoic acid was injected locally in 620 hamsters with buccal pouch cancers. The concentration of the agents were 250 ug/0.1 ml or 190 ug/0.1 ml. injected two times a week for four weeks. The following groups were used:

Group 1. DMBA
Group 2. DMBA + beta carotene (B.C.)
Group 3. DMBA + canthaxanthin (CTX)
Group 4. DMBA + 13-cis retinoic acid
Group 5. DMBA + saline
Group 6. Saline

Use of Liposomes to Regress Hamster Oral Tumor

Sixty Syrian hamsters were placed in six separate groups. The first three groups were treated with 0.5% DMBA for 16 weeks, three times per week, to induce oral squamous carcinoma. Groups 1 and 3 were injected locally at the tumor site with liposomes three times per week for five weeks. The liposomes were composed of phosphatidylcholine, phosphatidylserine, and phosphatidylethanolamine, in a ratio of 1:1:1 (a mixture of unilamelar and multilamelar vesicles, 70-300 nm). Group 1 animals were injected with the identical liposomes with beta carotene (389.2 ug/0.2 ml.). Additional groups were as follows: group 2, hamsters with DMBA induced tumors; group 4, normal non-tumor bearing, injected with liposomes of beta carotene three times per week for five weeks; group 5, normal, non-tumor bearing injected with liposomes (no beta carotene), three times per week for five weeks; and group 6, normal, non-tumor bearing hamsters.

Liposome Preparation

The liposome was formed from 50 mg. each of L-alpha phosphatidyl-L-serine, L-alpha phosphphatidylethanolamine from egg yolk, type III, and L-alpha phosphatidylcoline from egg yolk, type XI-E (Sigma Chemical, St. Louis, MO). The phosphatidylcholine in chloroform was dried with nitrogen gas. The 150 mgs were placed in 3.5 mls. of buffer which consisted of sodium phosphate (0.3 M) and potassium phosphate 0.3 M), at a pH of 7.5. This results in a final concentration of 40.0 mg./ml. The lipid mixture was then allowed to stand in a pyrex test tube (5-10 min). The constitutents were then vortex mixed for 10-15 min to dissolve all lipids, until a milky appearance was observed. Aliquots of 500 ul. were placed in test tubes of pyrex for sonication.

Beta Carotene-Liposome

The liposomes were sonicated for 10-15 min, then frozen in liquid nitrogen, then thawed. After thawing, 20 mg/0.5 ml., in phosphate buffer of beta carotene (Hoffman-La Roche), Nutley, NJ) was added to each tube and then sonicated again for 15 sec to 1 min. The liposomes were diluted to 10 ml. 10 ug/ml. of polylysine was added to facilitate centrifugation (1000 rpm). Liposomes of unincorporated beta carotene were decanted (23) and the liposomes were sized using polycarbonate filters (Nucleopore, Pleasanton, CA) (1.0 um). The liposomes were gently forced through the filter under pressure, producing mostly large unilamellar vesicles (70-300 nm), as determined in electronmicrographs. 0.1 mM of 6-carboxy-fluorescein was used to measure encapsulation efficiency (20-40%), and captured volume (7--12/ul./mg. of lipid).

Extraction of Beta Carotene from the Liposomes

After removal of 1 ml. from the liposome and beta carotene tubes, 1 ml. of 100% ethanol and 125 mg% of butanolyhydroxytoluene (BHT) were added to two volumes of hexane and diethyl ether (1:1), then shaken in a glass stoppered test tube and centrifuged (500 rpm) to separate the phases. The organic phase was then saved and the procedure repeated three times. The pooled extracts were then utilized for reverse phase high pressure liquid chromatography.

HPLC Determination

The amount of beta carotene in 1 ml. was determined by HPLC. The column was a 250X4.0 Bio-Sil, ODS-10 (Bio-Rad, Richmond, CA). The solvents used consisted of acetonitrile-methanol (85:15) and was then changed to hexane methanol (25:75) to remove the hydrocarbon fraction. Spectrophotometric analysis using the extinction coefficient (2200) was used to quantitate the amount of beta carotene present. The detector was set at a wavelength of 464-487 nm. The determination of the quantity of beta carotene per 1 ml. of liposome was calculated to be 1.94 mg. (389 ug./0.2 ml.; 19.4% retained in the liposomes).

Tumor Burden Analysis

The tumor burden was routinely assessed following the inversion of the hamster pouch and the counting of the number of visible tumors. The volume of the tumors was determined by assuming the tumors to be spherical. Therefore, the volume was calculated by the formula $4/3\pi r^3$ where r=diameter/2). Volume of each tumor was assessed. The mean tumor burden + S.D. for each group was also determined.

Immune Studies

Peritoneal Derived Macrophages. Resident peritoneal macrophages (RPM) were obtained from the peritoneum of the hamster following peritoneal lavage with (1X) phosphate buffered saline (PBS, 6 ml.), in a method previously described (Antoniades et al.,1984). The peritoneal exudate has been characterized following histochemical (nonspecific esterase, alpha napthyl esterase, acid phosphatase) (Sigma Chemical Co., St. Louis, MO) functional (Fc and C3 receptor) and immunohistochemical staining (tumor necrosis factor alpha (TNF-α) (Shklar and Schwartz, 1988).

Lymphocyte Isolation and Determination. Splenic lymphocytes were obtained following the formation of a single cell suspension and separation using nylon wool. This separation routinely provides a 95% enriched lymphocyte population. The lymphocytes were characterized by histochemistry (alpha napthyl esterase, negative) (Sigma) and immunohisto-chemically as follows: thy-1 positive (Thy 1.1 clone M5,5149 rat, antibody IgG2a) (1:20), Hybritech, Boehringer Mannhein Biochemical, IN), monoclonal antibody, MAb to TNF-beta, (1:20) rabbit anti-human, Endogen, Boston, MA), (MAb) positive 110, (Pan T) which specifies a T helper-inducer lymphocyte, that also induces interleukin 2 proliferation, and delayed type hypersensitivity, and MAb antiserum positive #38, an antiserum that specifies a cytotoxic population (Witte and Streilein, 1986; Witte et al., 1985), MAb #110 and #38 (mouse anti-hamster) were kindly provided by Dr. Pamela Witte, University of Texas, Dallas). The percentage of cytotoxic T lymphocytes (#38 positive cells) was presented as a fraction of the #110 positive cells. The presence of these cells in deparaffinized sections was determined using a peroxidase antiperoxidase immunohistochemical method (PAP). The controls included the replacement of the primary antibody with rabbit or hamster antiserum, and the substitution of a different antibody with the same isotype (rabbit anti-mouse IgG1, Cooper Biomedical, Mahvern, PA). An additional control for the activity of cytotoxic lymphocytes was the determination of cytotoxic lymphocytes following a cell mediated lympholysis assay (CML).

CML

Lymphocytes (10^5) from the spleen of the Syrian hamster were incubated with the lymphocytes from spleen of the Armenian hamster (5 X 10^5 cells, irradiated 2000 rads, cobalt) for 7 days. The T cells were incubated in T cell media (DMEM + 10% FBS + 2 ME (2-mercap-toethanol 10^{-5}M + 100 units Pen/Strept/ml. + T cell growth factor isolated from concanavalin A (Con A) stimulated BALB/C mouse splenic cells and then depleted by using a DE-50 column).

Cytotoxicity

Utilizing sodium chromate-radiolabelled ^{51}HCPC-1 tumor cells (10^5) and effectors (5 X 10^5) (RPM, and T and NK lymphocytes) (5:1 ratio of effectors to target). Effector cells were incubated for 24 hours then for an additional twenty hours (37°C) with the various agents. The wells were thoroughly washed and HCPC-1 tumor cells were added for an additional 20 hours. Specific lysis of the tumor cells was assessed as follows:

$$\% \text{ Specific Lysis} = \frac{100 \ (Re\text{-}Rs)}{100\text{-}Rs} \text{ where } \begin{array}{l} Re = \text{experimental release} \\ Rs = \text{spontaneous release} \end{array}$$

In general, the maximum release was 10,000 cpm, while spontaneous release was 10.0% of the maximum.

Microscopic Analysis

Light Microscopic Analysis. The hamster cheek pouch was removed and placed in 10% formalin for twenty-four hours. The paraffin embedded sections (6 μ), were stained with hematoxylin and eosin. These sections were then analyzed for histopathologic changes in the mucosa. Particular attention was given to the amount of dysplasia and invasion in the mucosa. An arbitrary scale was developed such that dysplasia was defined as demonstrating hyper-chromaticity, bizarre mitoses, nuclear-cytoplasmic reversal, pleomorphism, hyperplasia, and hyperkeratosis. A scale of 0-6 was utilized. The degree of invasion was recorded as follows: 1=loss of basement membrane, and slight penetration into the dermis, 2=dermal spread and into the muscle, 3=dermal invasion, and in all layers of the muscle, 4=in the dermis, through the muscle and into the lower dermis. Autopsies were performed to determine histopathologic changes in the liver, kidney, spleen, or lung. In addition to the analysis of the mucosa, a determination was made of the density and character of the inflammatory infiltrate. A scale was developed to determine the number of cells per 400X field, where 1=5-8 cells/field. Each analysis was performed by two pathologists, blind to the group and treatment.

Electron Microscopic Analysis. Pouches were fixed in cold Karnovsky's fixative and then processed for electron microscopy (Koller et al., 1973). Thin sections (600 Å) were cut with a diamond knife, stained with lead citrate and uranyl acetate after placement on a copper grid, and examined using a Zeiss EM 109 electron microscope. Thick (1.0) sections were stained with toluidine blue to assess the number of mast cells present in the tissue sections.

Statistical Analysis. Student's T test was determined with a confidence level of $p \leq .05$ for the assessment of significance.

RESULTS

Inhibition Studies

β-carotene was found to significantly inhibit the formation of DMBA-induced hamster buccal pouch tumor when applied topically on days alternate to the application of the carcinogen. An initial experiment with DMBA and β-carotene treated animals demonstrated fewer tumors with the β-carotene treatment than the DMBA controls and those tumors that did develop were smaller in size. Nine of the 10 animals in the DMBA group had grossly visible tumor, while only four of 10 animals in the β-carotene and DMBA treated group had gross tumor (Figures 1-2). Microscopically all tumors in all groups were histologically determined to be well differentiated squamous cell carcinoma. In general the pouches in the β-carotene treated group showed less dysplasia, (4.0, β-carotene, vs. 4.5-5.0 DMBA group) and less invasion into the dermis (2.5-3.0, β-carotene, vs. 4.0 DMBA group) (Figures 3-5). In an initiation and promotion study, the groups that received β-carotene at the time of initiation or promotion demonstrated fewer tumors than the DMBA treated group.

Prevention Study

Animals fed similar amounts of beta carotene, canthaxanthin, and a carotenoid mixture from an extract of algae were shown to have prevented or inhibited the gross development of squamous cell carcinoma induced in hamster buccal by DMBA. The tumor burden of the DMBA treated hamsters was 24 (mm³ X 10³). The tumor burden of the carotenoid treated group was 0, while the β-carotene and the CTX group showed a tumor burden of less than 1.0. The inhibition of tumor development in treated hamsters fed carotenoids was histologically unique. Microscopic areas of dysplasia, carcinoma in situ or early squamous cell carcinoma

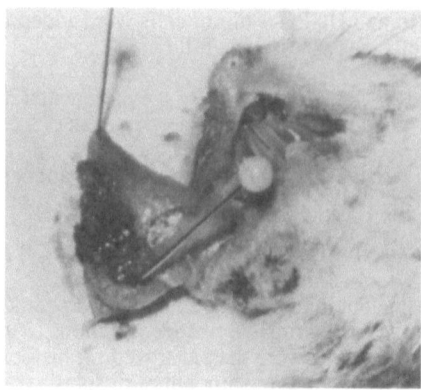

Figure 1. Syrian hamster treated with 0.5% DMBA for 16 weeks demonstrates the development of oral squamous cell carcinoma in the hamster cheek pouch.

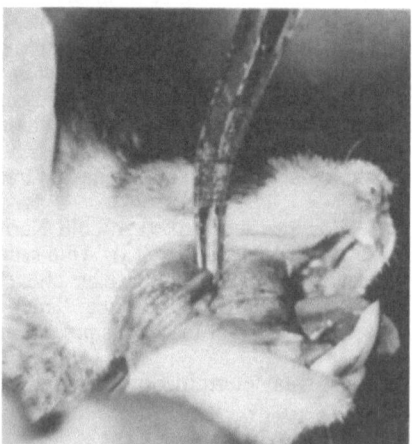

Figure 2. Syrian hamster treated with beta carotene on days alternate to the carcinogen DMBA. Note the tumors are smaller and fewer in number. In addition, a lesion of leukoplakia is seen in the hamster's cheek pouch (see arrow).

showed areas of tumor lysis and an inflammatory infiltrate consisting mainly of lymphocytes and histiocytes. Normal hamsters demonstrated 4-6 cells per 400 X microscopic field. These cells are generally lymphocytes, histiocytes, mast cells with a few PMNS (polymorphonuclear leucocytes). The lymphocyte population was found through immunohistochemistry staining to contain 5% peripheral helper inducer positive T cells, with only about one percent of the population identified as cytotoxic. In addition, only one percent of the inflammatory cells were histiocytes (macrophages) positive for TNF-alpha. Generally 40% of the spleen cells of the normal hamster was determined to be T lymphocytes, with 17% specified as cytotoxic T cells. In a cytotoxic assay using a tumor target, these cytotoxic cells only exhibited a 5.0% specific lysis of the tumor cells, in contrast the positive mitogen control, Con A, or cell inducer control, CMR, were 6-9 times more effective in lysing the tumor target. Macrophages (RPM) from the normal animal after incubation for 72 hours *in vitro* showed a small increase in cytotoxicity (14%), and only 9-18% of the spleen cells were positive for TNF-alpha, even though the level of production increased in the presence the tumor cell line HCPC-1 to 36%. Following DMBA treatment there was observed to be an increase in the inflammatory infiltrate. The number of cells per field was 8-12 at 400 X. The inflammatory cells were predominantly PMNs, with an increase of lymphocytes, histiocytes, and mast cells at dysplastic and carcinomatous sites. The level of dysplasia was designated as 6, while the level of invasion was 4, noting invasion through the muscle and into the underlying dermis. Specifically, the infiltrate was seen to contain more lymphocytes than the normal state, 18%, while only 5% were designated to be

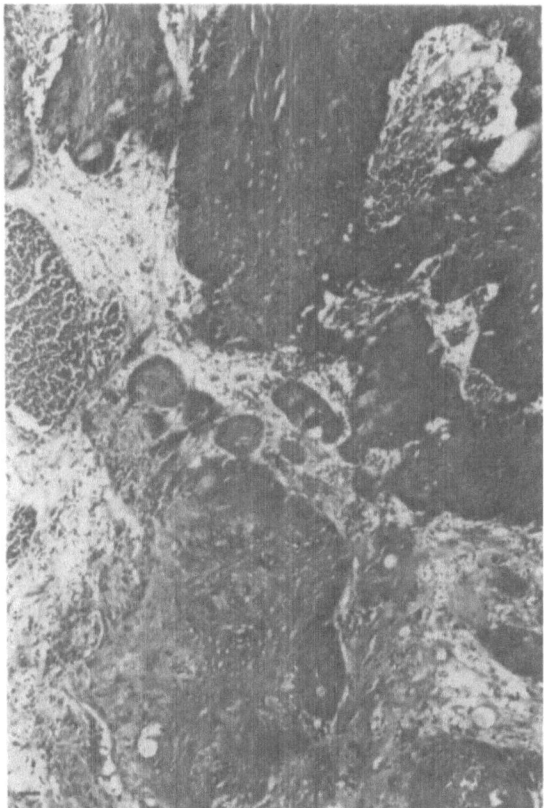

Figure 3. A photomicrograph of a squamous cell carcinoma of the hamster buccal mucosa induced by DMBA. The squamous cell carcinoma is seen invading deeply into the dermis.

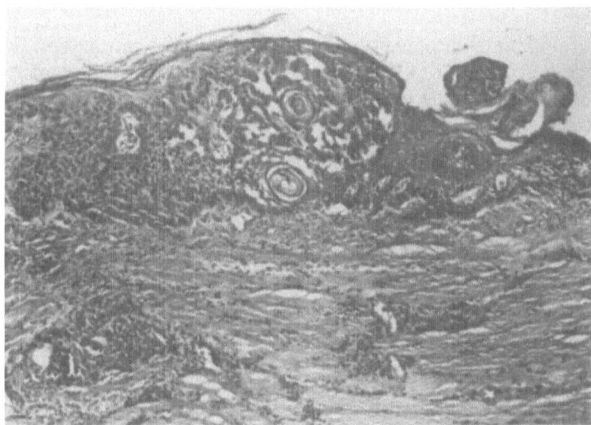

Figure 4. A photomicrograph of the oral mucosa of a hamster treated with beta carotene. Note that there are areas of dysplasia and carcinoma. In addition some areas of carcinoma are seen to be degenerating. An inflammatory infiltrate is seen throughout the dermis.

cytotoxic T lymphocytes. The cytotoxicity of these lymphocytes was remarkably reduced to 0.7%, while a similar assay for the RPM showed a similar reduction in specific lysis to 7%. On the other hand, there was an increase in TNF-alpha positive macrophages (10.0%) in the cancerous pouch. Upon treatment with the carotenoid mixture, B.C., or CTX, the number of

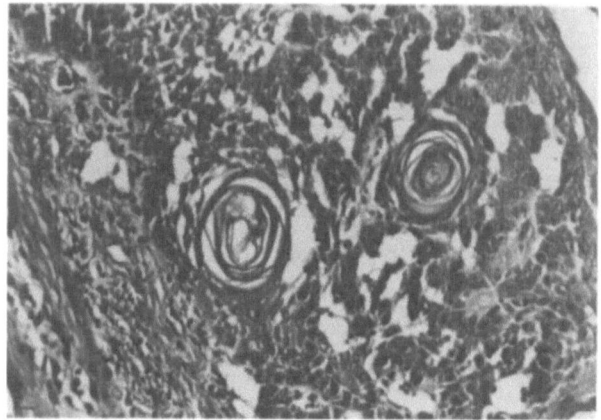

Figure 5. Higher magnification of area of tumor cells at various stages of cellular death.

cells per field increased to about 15-20. These cells included lymphocytes, histiocytes, mast cells, and PMNs. The level of dysplasia was approximated at 4.5 and the degree of invasiveness was generally, 1.5, which was associated with the loss of the basement membrane and invasion into the dermis. Characterization of T helper lymphocytes was 2-4 times the number of cells seen in the DMBA tumor bearing group. The level of cytotoxic lymphocytes was also increased 2-4 times the level of cytotoxic cells in DMBA tumor bearing groups. Carotenoid incubation in general, increased the number of identifiable cytotoxic T cells from the spleen, and the number of cytotoxic cells was also increased when incubated with the tumor cell line HCPC-1. The level of cytotoxicity following incubation with the carotenoids was demonstrated to be between 28-40%.

The number of histiocytes positive for TNF-alpha was 3 times the number in the tumor bearing animal. The cytotoxicity of the RPM was only slightly increased as compared to RPM from tumor bearing hamsters, RPM upon incubation for 24 hours with the carotenoids, the number of TNF-alpha positive cells increased 7-8 times the number of TNF positive cells in the normal hamster.

Hamsters treated with beta carotene on days alternate to the carcinogen DMBA were utilized to ascertain the percentage of GGT positive areas for each of the groups. GGT levels/whole mount, for beta carotene treated animals was 2.5% while DMBA treated hamsters demonstrated an increase to 7.5%.

Regression Studies

In initial regression experiments beta carotene applied either topically or injected locally at tumor sites was observed to produce a 60% reduction in tumor. In another series beta carotene injected locally to the oral tumor of the hamster, resulted in 20% of the hamsters exhibiting complete regression of carcinoma, and 80% of the other animals in the group exhibited partial regression. In this study CTX demonstrated a 15% complete regression of oral tumor, while 85% showed a partial regression of tumor. The vitamin A derivative, 13-cis retinoic acid, showed no complete regression while 70% of the animals were noted to have a partial response. In a third study, there was observed that the beta carotene group had 52 times less tumor than the solely treated tumor bearing group. CTX treated hamsters showed a 16 times less tumor than sham treated controls. On the other hand, 13-cis retinoic acid was observed to have an increased tumor burden which was 1.2 times greater than the tumor controls. This trend of tumor reduction following local injection of B-carotene, and CTX was continued in a fourth study (Figures 6-9). Histopathologic evidence indicated the lysis of tumor cells and the apparent lack of effect on adjacent normal oral mucosa. In addition there was observed to be an increase in the inflammatory infiltrate (Figure 10). The infiltrate was characterized as a mixed cellular infiltrate, which consisted of lymphocytes, histiocytes, mast cells, and some PMNs. In addition, the level of TNF-alpha was found to be between 67-87% of the cells per mm^2 field (400X) as compared to the control tumor group.

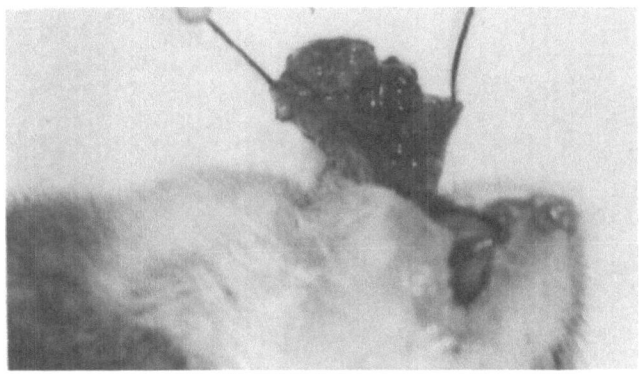

Figure 6. Photograph of sham injected DMBA induced oral carcinoma of the Syrian hamster.

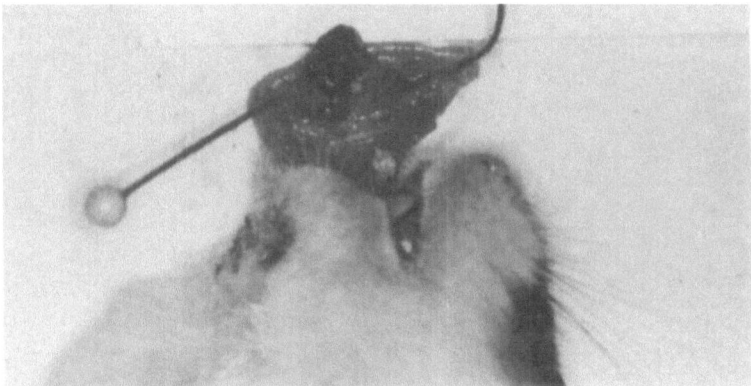

Figure 7. Photograph of hamster oral carcinoma injected with 13-cis retinoic acid. Note the oral carcinoma has not undergone tumor destruction or inhibition of tumor growth as compared to the sham control.

Figure 8. Following the injection of beta carotene a significant reduction in tumor burden is seen as compared to sham injected control.

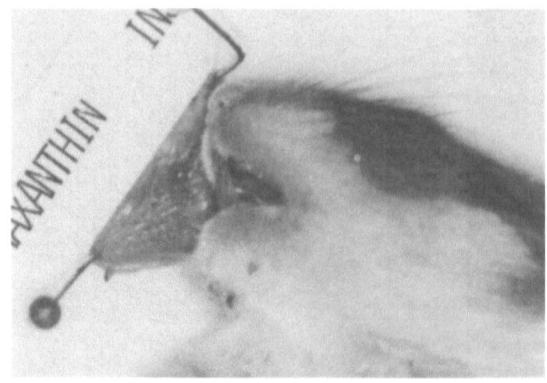

Figure 9. The administration of canthaxanthin is seen to produce a reduction in oral carcinoma similar to beta carotene treatment.

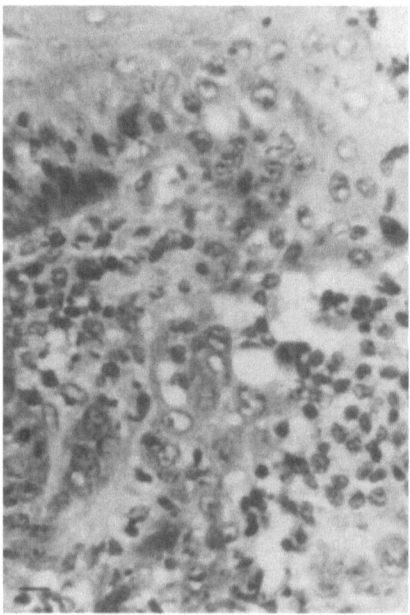

Figure 10. Photomicroscopic picture of the tumor cell lysis and cellular infiltrate seen following beta carotene injection.

Liposome Regression Study

Utilizing liposomes of beta carotene there was noted to be a significant ($p \leq .001$) reduction in tumor burden in those hamsters that were injected locally. The electron-micrographic analysis demonstrated an endocytosis of liposomes of beta carotene and their accumulation in squamous cell carcinomas, without a corresponding change in the normal mucosa. The accumulation of liposomes resulted in the lysis of the tumor cells. In addition, the swelling or rupture of mitochondria, lysosomes, golgi apparatus, and endoplasmic reticulum was observed (Figures 11-14). Histologic analysis of the cellular inflammatory infiltrate indicated that the β-carotene liposome injections resulted in an increase of mononuclear cells (lymphocytes+histiocytes) (60-65%), PMNs (10-15%), mast cells (10-15%), and endothelial lined vascular spaces (2-5%). These cellular values were in contrast to the DMBA control values which indicated a reduction of mononuclear cells, a significant increase of PMNs ($p \leq .001$), and an increase in endothelial lined vascular spaces. The percentage of cellular change for each group was compared to normal hamster cellular levels.

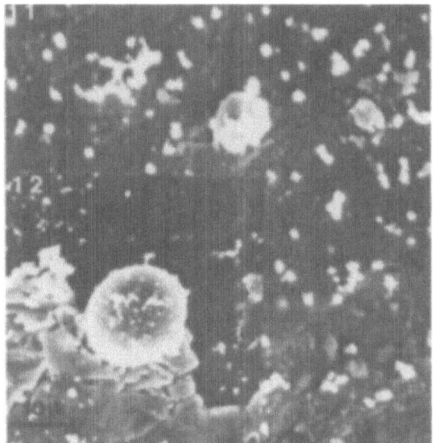

Figure 11. Scanning electron image of liposomes of beta carotene.
Figure 12. Note higher magnification of scanning electron microscopic image of the beta carotene liposome insert.

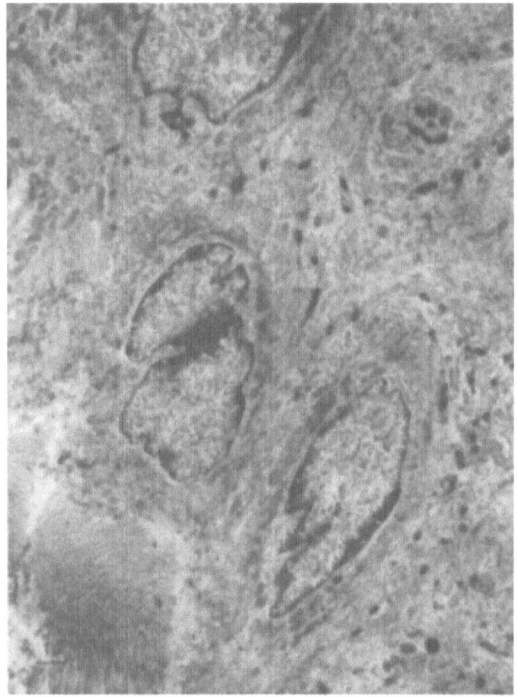

Figure 13.   Electron microscopic picture of normal oral mucosa of the hamster pouch treated with beta carotene liposomes. Note there is no apparent liposomes in the normal mucosa.

DISCUSSION

Beta carotene administration has resulted in the prevention, inhibition and regression of oral squamous cell carcinoma in the hamster tumor model. In addition, histopathologic analysis of the cellular infiltrate associated with the inhibition and regression demonstrated a unique increase in the number of cellular infiltrates, and the type of inflammatory cells. Specifically, both cytotoxic lymphocytes and macrophages were present and appeared to be associated with the depression in tumor growth (Schwartz et al., 1986).

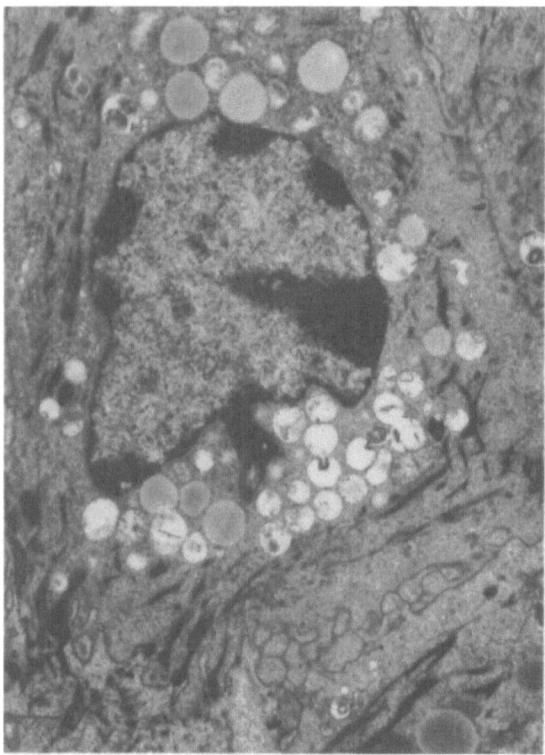

Figure 14.   Beta carotene liposomes are seen accumulating in an oral carcinoma tumor cell. The accumulation of liposomes apparently results in the degeneration of the cytoplasmic organelles of the tumor cell.

The selective endocytosis of beta carotene liposomes and the lack of uptake by normal keratinocytes provides the indication that tumor cells have differences in membrane associated phospholipids and proteins (Porter and Wagner, 1986). Therefore, there was a difference in the ability of liposomes to adhere to the cell membrane and induce the endocytosis process (Martin and MacDonald, 1976). This process involves the displacement of membrane receptors, membrane associated enzyme-substrate complexes, changes in the biochemistry of the membrane, and the action of microfilaments to move the endocytic vesicle inward, and deeper into the cytoplasm (Mayhew, et al., 1980; Szoka et al., 1981). The induction of beta carotene, specifically, in the form of a liposome may induce the autoperoxidation of the membrane (Pagano and Huang, 1975). This would result in the release of reactive oxygen anions, or oxygen metabolites (Cerutti, 1985; Begin et al., 1988). These reactive products are controlled in normal cells with superoxide dismutases (SOD) (Oberley and Buettner, 1979). Cu-Zn superoxide dismutase which has been associated with the cellular membrane-cytosol, has been shown to be reduced in solid tumors. Another dismutase, Mn-superoxide dismutase, has been linked to mitochondrial membrane, the scavenging of potential toxic superoxide radicals, and has been found to be reduced in solid tumors such as carcinomas (Wong and Goeddel, 1988). Following the introduction of the beta carotene liposomes into the tumor cells the swelling of the mitochondria and disruption of the lysosomes was evident. This organelle disorganization may result in the reduction in Mn-SOD and the release of phospholipases, and phosphatases into the cytosol, further altering the metabolic state of the tumor cell (Valdimirov, 1986).

It is interesting to note that TNF-alpha can induce the mRNA and presumably the active enzyme (Mn-SOD) *in vivo* and in cell lines (Wong and Goeddel, 1988). TNF-alpha as indicated previously has been determined to be produced at the site of tumor prevention and regression (Shklar and Schwartz, 1988). TNF-alpha has been shown to induce *in vivo* tumor necrosis, and has demonstrated various capacities to regulate immune cytotoxicity indirectly through the action of gamma interferon, and IL-1 (Talmadge et al., 1987). Therefore TNF-alpha induction of Mn-SOD activity may be blocked by the action of beta carotene, other carotenoids, or terpenoids. This inhibition would possibly shift the TNF-alpha activity to lysis of the tumor cell. Other inhibitors of oxidative metabolites such as catalases, and glutathione peroxidases, would also be important in controlling cellular-organelle damage (Meister, 1983). In terms of the glutathione (GSH) pathway, various studies have indicated that elevated reduced GSH levels are associated with inhibition of carcinogenesis (Rothstein and Slaga, 1988). The level of GGT (gamma-glutamyl-transpeptidase), an enzyme of importance in the production of the oxidative form of GSH and gamma-glutamyl cysteine formation was found to be reduced following beta carotene inhibition of carcinogenesis (Suda et al., 1987), perhaps indicating an accumulation of reduced GSH. In human patients with a deficiency of GGT, reduced GSH levels are high enough to be excreted into the serum and the urine (Carmichael et al., 1988).

The stages of beta carotene action in the process of carcinogenesis has not been fully determined. *In vivo* studies indicate that beta carotene inhibits the promotion stage of tumor development (Nishino et al., 1988; Mathews-Roth and Krinksy, 1984). There was also observed in one of the vivo studies an inhibition in DMBA initiation (Suda et al., 1986). This finding has been partially substantiated by an *in vitro* study, in which beta carotene was noted to inhibit transformation induced by 3-methylcholanthrene in a fibroblast cell line (10T1/2 cells) (Pung et al., 1988). In addition, in a similar type of study, it was noted that the inhibition of transformation by beta carotene was a carotenoid effect, and was not due to conversion into a vitamin A retinoid derivative (Rundhaug et al., 1988). This evidence agrees well with the *in vivo* data presented here. In both *in vivo* and *in vitro* studies the carotenoid, CTX, was seen to exert a potent inhibitory-preventive response. CTX, as noted previously, does not convert to vitamin A.

The use of DMBA in the hamster produces a profound immune cellular depression, of lymphocytic effectors (Elmets et al., 1988). The serum picture as well as the histologic analysis of DMBA induced oral carcinoma of the hamster indicated a significant increase in PMNs and a decrease in the mononuclear cells (histiocytes and lymphocytes). The administration of beta carotene or other carotenoids such as CTX appeared to reverse this cellular picture. The serum and the histologic analysis of the hamster cheek pouch indicated an increase in mononuclear cells, and mast cells and fewer PMNs. In addition it was observed that there was a decrease in the vascularization of the tumors. These general observations were consistent in inhibition, as well as in regression studies, although the response was most striking in the prevention-inhibition studies. The increase and the selective accumulation of these cells was probably associated with the development of chemotactic gradients, changes in cellular adherence, and increased cellular interactions. The details of this response remain to be determined. The systemic increase in immune reactivity has been noted through the use of cytotoxicity assays (Schwartz et al., 1986, and Tomita et al., 1987). Following DMBA tumor induction there was observed to be present a reduction in cytotoxicity by RPM following incubation with the cloned tumor cell line HCPC-1 (Shklar and Schwartz, 1988). Additional studies following the inoculation of this cell line into the peritoneum of the hamster indicated that the carcinoma cells were inhibiting the cytotoxic response by the RPM. This inhibition could be reduced if the hamster was administered beta carotene submucosally into the buccal pouch or into the peritoneum (unpublished data). It is therefore plausible that beta carotene, other carotenoids, as well as other antioxidants, such as vitamin E, can enhance the antitumor activity of immune effectors, by increasing their efficiency in producing reactive peroxide and oxygen metabolites, by increasing the antioxidant activity in these activated immune cells (Kitagawa and Johnston, 1986). In addition as the cells become activated the interaction between cell types, particularly lymphocyte-monocyte interactions which are associated with the cascade of interleukin-cytokine production would also presumably be increased (Goldyne and Stobo, 1981). Finally the endocytosis process by the tumor cells, may alter the antigenicity of the tumor membrane, therefore enhancing immune recognition and reactivity to the tumor cells. Future studies will investigate this immune response mechanism, as it relates to carotenoid triggering and targeting to the tumor site.

REFERENCES

Alam, B. S., Alam, S. Q., Weir, J. C., Jr., and Gelson, W. A., 1984, Chemopreventative effect of B-carotene and 13-cis retinoic acid in salivary gland tumors, **Nut. and Cancer,** 6:4.

Antoniades, D. Z., Schwartz, J. L., and Shklar, G., 1984, The effect of chemically induced oral carcinoma on peritoneal macrophages, **J. Clin. Lab. Immunol.,** 14:17.

Begin, M. E., Ells, G., and Horrobin, D. F., 1987, Polyunsaturated fatty acid-induced cytotoxicity against tumor cells and its relationship to lipid peroxidation, **J. Natl. Cancer Inst.,** 80:188.

Bendich, A., and Shapiro, S. S., 1986, Effect of B-carotene and canthaxanthin on the immune responses of the rat, **J. Nutr.,** 116:2259.

Cerutti, P. A., 1985, Prooxidant states and tumor promotion, **Science,** 227:375.

Elemets, C. A., Khan, W. A., Klemme, J. C., and Mukhtar, H., 1988, Impaired immunologic surveillance by 7,12 dimethylbenz(a) anthracene augments its skin tumorigenicity in C3H mice, **Biochem. Biophys. Res. Commun.,** 151:148.

Goldyne, M. E., and Stobo, J. D., 1981, Immunoregulatory role of prostaglandins and related lipids, **Crc. Crit. Rev. Immuno.,** 3:189.

Kitagawa, S., and Johnston, R. B., Jr., 1986, Deactivation of the respiratory burst in activated macrophages: Evidence for alternation of signal transduction, **J. Immunol.,** 136:2605.

Koller, T. M., Baer, M., Muhler, M., Kuhlethaier, K., 1973, Electron microscopy of selectively stained molecules, *in*: Principles and techniques of electron microscopy," M. A. Hayat, ed., Van Nostrand, New York.

Martin, F. J., and MacDonald, R. C., 1976, Lipid-vesicle-cell interactions. II. Induction of cell fusion, **J. Cell. Biol.,** 70:506.

Mathews-Roth, M. M., 1982, Antitumor activity of B-carotene canthaxanthin and phytoene, **Oncology,** 39:33.

Mathews-Roth, M. M., and Krinsky, N. I., 1984, Effect of dietary fat level on UV-B induced skin tumors, and antitumor of B-carotene, **Photochem Photobiology** 40:671.

Mayhew, E., Gotfredsen, C., Schneider, Y. J., and Trouet, A., 1980, Interaction of liposomes with cultured cells, effect of serum, **Biochem. Pharm.,** 29:877.

Meister, A., 1983, Selective modification of glutathione metabolism, **Science,** 220:472.

Menkes, M. S., Comstock, G. W., Vuilleumer, J. P., Helsing, K. J., Rider, A. A., Brookmeyer, R., 1986, Serum beta carotene, vitamin A and E, selenium, and the risk of lung cancer, **New Engl. J. Med.,** 315:1250.

Nishino, H., Takayasu, J., and Hasegawa, T., 1988, Biological activities of natural carotenes, **4th Internat. Congress of Cell Biol.-Montreal,** P1., 2.40a.

Oberley, L. W., and Beutlner, G. R., 1979, Role of superoxide dismutase in cancer: A review, **Cancer Res.,** 39:1141.

Odukoya, O., Schwartz, J. L., Weichselbaum, R., and Shklar, G., 1983, An epidermoid carcinoma cell line derived from hamster 7,12 dimethylbenz(a)anthracene induced buccal pouch tumor, **J. Natl. Cancer Inst.,** 71:1253.

Odukoya, O., and Shklar, G., 1984, Initiation and promotion in experimental oral carcinogenesis, **Oral Surg.,** 58:315.

Ohno, Y., Yoshida, O., and Oishi, K., 1988, Dietary B-carotene and cancer of the prostate: A case control study in Kyoto, Japan, **Cancer Res.,** 48:1331.

Pagano, R. E., and Huang, L., 1975, Interaction of phospholipid vesicles with cultured mammalian cells, **J. Cell Biol.,** 67:49.

Porter, N. A., and Wagner, C. R., 1986, Phospholipid autoxidation, *in*: "Advances in free radical biology and medicine," vol. 2, W. A. Pryor, ed., Pergaman Press, London.

Pung, A., Rundhaug, J. E., Yoshizawa, C. N., and Bertram, J. S., 1988, B-carotene, and canthaxanthin inhibit chemically- and physically-induced neoplastic transformation in 10T1/2 cells, **Carcinogenesis,** 9:1533.

Rettura, G., Stratford, F., Levenson, S. M., and Seifter, E., 1982, Prophylactic and therapeutic action of supplemental B-carotene in mice inoculated with C3HBA adenocarcinoma cells. Lack of therapeutic action of supplemental ascorbic acid, **J. Natl. Cancer Inst.,** 69:73.

Rhodes, J., 1983, Human interferon action: reciprocal regulation by retinoic acid and B-carotene, **J. Natl. Cancer Inst.,** 70:833.

Rhodes, J., Polwan, P., Bishop, M., Lipscomb, D., 1981, Human macrophages function in cancer: Systemic and local changes detected by an assay for Fc receptor expression, **J. Natl. Cancer Inst.,** 66:423.

Rothstein, J. B., and Slaga, T. J., 1988, Effect of exogenous glutathione on tumor progression in the murine skin multistage carcinogenesis model, **Carcinogenesis,** 9:1547.

Rutenberg, A. A., Kim, H., Fishbein, J., Harper, J., Wasserkrug, H., Seligman, A., 1969, Histochemical and ultrastructural demonstration of gamma-glutamyl transpeptidase activity, **J. Histochem. Cytochem.,** 17:517.

Schwartz, J. L., and Shklar, G., 1988, Regression of experimental oral carcinoma by local injection of beta carotene and canthaxanthin, **Nutr. and Cancer,** 11:35.

Schwartz, J. L., and Shklar, G., 1987, Regression of experimental oral cancer by beta carotene and algae extract, **J. Oral MaxilloFac. Surg.,** 45:510.

Schwartz, J. L., Suda, D., and Light, G., 1986, Beta carotene is associated with the regression of hamster pouch carcinoma and the induction of tumor necrosis factor in macrophages, **Biochem. Biophys. Res. Commun.,** 136:1130.

Seifter, E., Rettura, G., and Padawer, J., 1984, Regression of C3HBA mouse tumor due to X-ray therapy combined with supplemental B-carotene or vitamin A, **J. Natl. Cancer Inst.,** 71:409.

Shklar, G., 1972, Experimental oral pathology in the Syrian hamster, **Prog. Exp. Tumor Res.,** 16:518.

Shklar, G., and Schwartz, J. L., 1988, Tumor necrosis factor in experimental cancer regression with alphatocopherol, beta carotene, canthaxanthin, algae extract, **Eur. J. Cancer Clin. Oncol.,** 24:839.

Slaga, T. J., Klein-Szanto, A. J. P., Triplett, L L., Yotti, L. P., and Trosko, J. E., 1981, Skin tumor promotion activity of benzoyl peroxide, a widely used free radical-generating compound, **Science,** 213:1023.

Solt, D. B., 1981, Localization of gamma-glutamyl transpeptidase in hamster buccal pouch epithelium treated with 7,12 dimethylbenz(a) anthracene-treated hamster buccal pouch, **Cancer Res.,** 42:285.

Solt, D. B., and Shklar G., 1982, Rapid induction of - glutamyl transpeptidase-rich intraepithelial clones in 7,12 dimethylbenz(a)anthracene-treated hamster buccal pouch, **Cancer Res.,** 42:285.

Stich, H. F., and Dunn, B. P., 1986, Relationship between cellular levels of beta carotene and sensitivity to genotoxic agents, **Int. J. Cancer,** 38:713.

Stich, H. F., Hornby, A. P., and Dunn, B. P., 1986, Beta carotene levels in exfoliated mucosa cells of population groups at low and elevated risk for oral cancer, **Int. J. Cancer,** 37:389.

Stich, H. F., Rosin, M. P., Hornby, A. P., Mathews, B., Sarkaranarayanan, R., and Nair, M. K., 1988, Remission of oral leukoplakias and micronuclei in tobacco/betel quid chewers treated with beta carotene and with beta carotene plus vitamin A, **Int. J. Cancer,** 42:195-199.

Suda, D., Schwartz, J. L., and Shklar, G., 1986, Inhibition of experimental oral carcinogenesis by topical beta carotene, **Carcinogenesis,** 7:711.

Talmadge, J. E., Phillips, H., Schneider, M., Rowe, T., Pennington, R., Bowersox, O., and Lenz, B., 1988, Immunomodulatory properties of recombinant murine and human tumor necrosis factor, **Cancer Res.,** 48:544.

Tomita, Y., Himeno, K., Nomoto, K., Endo, H., and Hirohata, T., 1987, Augmentation of tumor immunity against syngeneic tumors in mice by β-carotene, **J. Natl. Cancer Inst.,** 78:679.

Wald, N. J., 1982, Vitamin A and cancer in humans, *in:* "Disease and the Environment," R. Rees and H. J. Purcell, eds., John Wiley, Chichester.

Wang, G. H. W., and Goeddel, D. V., 1988, Induction of manganous superoxide dismutase by tumor necrosis factor: Possible protective mechanism, **Science,** 242:941.

Witte, P. L., Stein-Streilein, J., and Streilein, J. W., 1985, Description of phenotypically distinct T lymphocyte subsets which mediate helper, DTH and cytotoxic function in the Syrian hamster, **J. of Immunol.,** 134:2408.

Witte, P. L., and Streilein, J. W., 1986, Development and ontogeny of hamster cell subpopulations, **J. of Immunol.,** 137:45.

STUDIES ON MEMBRANE LIPID PEROXIDATION IN OMEGA-3 FATTY ACID-FED

AUTOIMMUNE MICE: EFFECT OF VITAMIN E SUPPLEMENTATION

Serge Laganiere, Byung P. Yu and Gabriel Fernandes

Department of Medicine and Physiology
The University of Texas Health Science Center at San Antonio

ABSTRACT

Enzyme-dependent and non-enzymatic *in vitro* lipid peroxidation was studied in autoimmune prone B/W mice fed diets containing high levels of dietary corn oil (CO) or menhaden fish oil (FO) as lipid source since weaning. Lipid analysis revealed that FO-fed mouse liver mitochondrial and microsomal membrane fractions incorporated 20:5 ω 3 and 22:6 ω 3 in replacement of 18:2ω6 and 20:4ω6 found in corn oil (CO) fed control animals reflecting the composition of the dietary oils. Lower concentrations of vitamin E were found in the FO-fed mouse membranes and serum than those of CO-fed mice when diets were supplemented with a standard 75 I.U. α-tocopheryl acetate/kg diet. The rate and extent of membrane lipid peroxidation was greatly increased in FO-fed, vitamin-E-depleted membranes. Full repletion of membrane vitamin E levels by supplementation with 500 I.U./kg of FO diet for 30 days significantly decreased lipid peroxidation and showed that in FO-fed mice, membrane peroxidation is inversely proportional to vitamin E content. However, due to a lower ratio of vitamin E and highly unsaturated fatty acids, FO-fed mouse membranes were more sensitive to pro-oxidant stimulus than were those from CO-fed mice. These findings illustrate the action of vitamin E against membrane lipid peroxidation and stress the importance of adequate supplementation of antioxidant with high omega-3 fatty acids intake.

INTRODUCTION

The lower incidences of thrombosis and ischemic heart disease in Greenland Eskimos were attributed to their high intake of omega-3 fatty acids derived from marine products (Bang and Dyerberg, 1980). These findings have stimulated a great deal of research interest to further characterize the functional properties of dietary fish oil (FO). Studies in human subjects indicated substantial lowering of blood lipids and lipoproteins (Phillipson et al., 1985) and reduced platelet aggregation and thromboxane formation (Siess et al., 1980). Furthermore, FO-ω3 fatty acids may correct immune dysfunction. For example, the decrease arthritis susceptibility in mice (Leslie et al., 1985), reduce markedly the severity of glomerulonephritis in (NZB x NZW) (B/W) autoimmune mice (Prickett et al., 1981; Fernandes, 1987) and improve the clinical symptoms of rheumatoid arthritis (Kremer et al., 1986).

The mechanisms underlying the action of FO on lymphoid cells are not yet fully known but the C20 and C22-3 polyunsaturated fatty acids (PUFA) of FO are incorporated into cell membranes (Stubbs and Smith, 1984), replacing major PUFA like arachidonic acid and linoleic acid with eicosapentaenoic (20:5) (EPA) and docosahexaenoic (22:6) (DHA) acids. The incorporation of EPA and DHA fatty acids containing numerous methylene interrupted double bonds may alter the membrane bilayer fluidity (Stubbs and Smith, 1984) and could increase membrane lipid peroxidizability (Holman, 1954). Therefore, antioxidant requirements for maintaining membrane integrity may increase when a highly enriched FO diet is consumed. We addressed this particular question in B/W mice and now report that chronic intake of a high FO-enriched diet decreases vitamin E levels in membranes and serum, and increases liver mitochondrial and microsomal lipid peroxidation. However, increased levels of vitamin E supplementation reverses most of these adverse effects.

MATERIALS AND METHODS

Mice

Inbred male (NZB X NZW)F$_1$ (B/W) mice of 4 weeks of age were obtained from Jackson Laboratories (Bar Harbor, ME). The mice were housed under standard animal care conditions and were fed with Purina Laboratory Chow and maintained in controlled humidity and temperature with alternating light and dark cycles. Animals were started at 8 weeks of age on a semipurified diet as described below and were used at 5 months of age for experiments.

Diets

The basic formula of the semi-purified diet was (w/w): 20% casein, 25% dextrose, 25% starch, 20% fat (corn oil or menhaden oil), 3.5% cellulose, 3.5% mineral mixture, 1.5% vitamin mixture, 0.3% dL-methionine, 0.2% choline and 1% agar. This represents a standard high-fat diet, which has been shown in many studies to promote neoplastic and autoimmune disease. The proportion of high fat approximates that found in the basic Western diet. The standard vitamin mixture (ICN, Cleveland, OH) provided 75 I.U./kg diet of dl-α-tocopheryl acetate. In the present experiment, another group of age-matched, menhaden-oil-fed mice was supplemented with α-tocopheryl acetate to a total of 500 I.U./kg diet and fed for one month to compare vit E levels and *in vitro* lipid peroxidation to those found in animals maintained on 75 I.U./kg diet.

Membrane Preparations

Mice were anesthetized with ether and exsanguinated. The liver tissue of 2 mice was pooled for each preparation and mitochondrial and microsomal membrane fractions were isolated by differential centrifugation as described previously (Laganiere and Yu, 1987)

In Vitro Peroxidation

Membrane peroxidation was assessed by non-enzymatic and enzyme-dependent *in vitro* lipid peroxidation (LPO) as described (Laganiere and Yu, 1987) with the following modification. Non-enzymatic LPO was carried out in the presence of 5 μM each of ferrous sulfate and ferric chloride. Extent of LPO was followed by a modification of the thiobarbituric acid (TBA) assay of Ohkawa et al. (1979). TBA (0.6 g) was dissolved in 1N NaOH, then titrated to pH 3.5 with 7% perchloric acid (1.2% w/v final concentration). With this procedure, a 10 minute centrifugation at 2000 X g replaces the lengthy organic extraction of the pigment (Ohkawa et al., 1979) and increases the sensitivity. Malonaldehyde (MDA) bisdiethyl acetal (Aldrich Chemicals, Milwaukee, WI) was used as a standard).

Vitamin E

Membrane and serum vit E levels were determined by a spectrofluorometric procedure (Taylor et al., 1976), α-tocopherol (Sigma Chemical Co., St. Louis, MO) was used as a standard.

Fatty Acid Analysis

Membrane lipids were extracted in 2:1 chloroform/methanol (Folch et al., 1957) and fatty esters were derivatized with methanolic-HCl. Fatty acid analyses were conducted by gas-liquid chromatography using a fused silica capillary column as described (Laganiere and Yu, 1987). Identity of each fatty acid was verified by mass spectrometry (courtesy of Dr. Susan Weintraub, Dept. of Pathology, UTHSCSA). Peroxidizability coefficients were calculated as described by Witting and Horwitt (1964).

Data Analysis

All data were composed by multiple analysis of variance and with Duncan's multiple-range test analyzed by the Student's t-test.

RESULTS

Concentrations of the major fatty acids in the menhaden oil that was used in this study were: 16:0, 19%; 16:1, 12%; 18:1, 15%; 18:2, 5%; 20:5 (EPA), 15%; 22:6 (DHA), 11%. Conversely, in corn oil (CO) 18:1 content was 27% and 18:2 was 58%.

When both 20% fat diets were supplemented with a standard 75 I.U. α-tocopherol/kg diet, serum and vit-E contents of the liver membrane vit E contents were significantly lower in the FO- than in the CO-fed group ($p < 0.005$). Table 1 shows that serum and membrane vit E levels were depressed to approximately one quarter their former levels after 4 months of chronic FO feeding. Data also show that membrane content was fully repleted by increasing the α-tocopherol ration to 500 I.U./kg diet and serum vit E was increased 2.5-fold after 30 days of supplementation.

Table 1. **Effect of dietary oil and a-tocopherol supplementation upon concentrations of serum and liver membrane vitamin E.**

	Serum (mg/mL)	Liver	
		Mitochondria (mg/g prot)	Microsomes (mg/g prot)
CO + 75 I.U.	8.38[a](11)	282.1[b](5)	542.4[c](5)
	±0.06	±30	±56
FO + 75 I.U.	1.95	74.9(4)	161.9(4)
	±0.2	±11	115
FO + 500 I.U.	5.09[a](9)	297.8[b](3)	448.5[c](3)
	±0.5	±32	±69

Values are expressed at mean ± SEM with number of preparations in parenthesis. CO vs FO + 500 IU: [a]$p < 0.01$, [b,c]non-significant. CO vs FO + 75 IU: $p < 0.005$ for each parameter:

The extent of membrane lipoperoxidation was assessed by exposing liver mitochrondrial and microsomal membrane fractions to pro-oxidant conditions for induction of thiobarbituric acid reactive substances (generation of MDA.) Analysis of the time course of non-enzymatic (Fig. 1) or enzyme-dependent LPO (Fig. 2) of CO-fed mouse microsomal fractions showed a period of latency at between 5 and 10 minutes incubation. Such a lag in the initiation of LPO was not observed in mitochrondria. In contrast, there was no microsomal lag period and peroxidation was much more extensive in vit E depleted membrane fractions from FO-fed mice. In fact, after only 5 minutes incubation, peroxidation was comparable to the levels observed at 60 minutes in the CO-fed group. In vit E repleted membrane fractions, rates of LPO were clearly decreased, especially in non-enzymatic LPO, as illustrated by the lower accumulations of MDA, although microsomal lag period was not restored. Figs. 1 and 2 indicate that vit E repleted membrane fractions from the vit E supplemented, FO-fed mice remained more peroxidizable than CO-fed mouse membrane fractions since, with time, accumulation of MDA tended to approach that of the vitamin E depleted membrane fractions especially in enzyme-dependent LPO. Chronic intake of a FO diet did not modify the membrane protein recovery or the NADPH cytochrome c (P-450) reductase specific activity (results not shown).

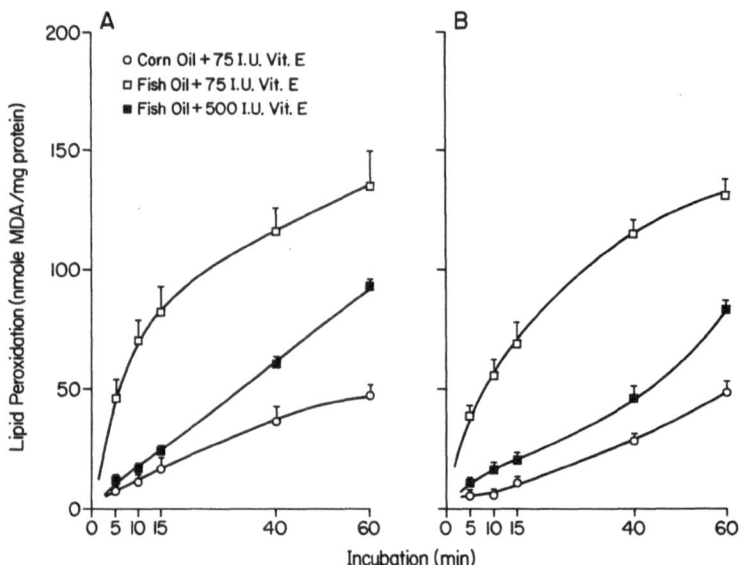

Fig. 1. *In vitro* non-enzymatic lipid peroxidation. A) Mitochondrial membranes. B) Microsomal membranes. Incubation system contained 37.5 mM Tris (pH 7.4), 112 mM KCl, 5 mM FeCl3. Reaction was started by adding ascorbic acid at 0.5 mM final concentration. Peroxidation took place at $30^{\circ}C$ with vigorous shaking. Each point represents the mean ± SEM of three separate experiments.

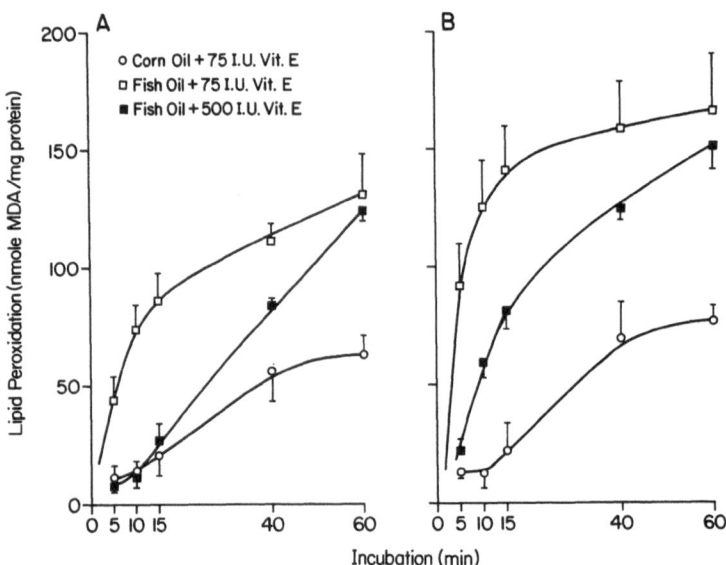

Fig. 2. *In vitro* enzyme-dependent lipid peroxidation. A) Mitochrondrial membranes. B) Microsomal membranes. Incubation system contained 37.5 mM Tris (pH 7.4), 112 mM KCl, 0.1 mM FeSO4 complexed with 5 mM ADP. Reaction was started by adding 1 mM NADPH. Peroxidation took place at $30^{\circ}C$ with vigorous shaking. Each point represents the mean ± SEM of three separate experiments.

Analysis of liver mitochrondrial and microsomal fraction lipids confirmed that membranes incorporated the respective dietary fatty acids. Table 2 reveals that CO-fed mouse membrane fractions contained 36% linoleic plus arachidonic acid as compared to only 12-14% in FO-fed animals. Interestingly, the arachidonic/linoleic ratio was closely maintained at about 0.8 in all groups studied. EPA and DHA content was 11% in CO-fed mouse membrane fractions and increased to nearly 30% in the FO-fed group membrane fractions. Computation of the resulting membrane peroxidizability coefficient (Pc) as described in Table 2 showed that Pc's were greatly increased in the FO-fed groups (p < 0.005). Despite the striking changes in fatty acid profiles, the double-bond index remained constant. This was partly due to a significant increase in palmitic acid contents in FO-fed mouse membrane fractions (p < 0.01). Table 2 also shows that the membrane fatty acid profiles of the two FO-fed groups (75 vs 500 I.U. α-tocopheryl acetate/kg diet) were not significantly different.

DISCUSSION

The present studies indicate that membrane and serum vit E were reduced in FO-fed (NZB X NZW) mice supplemented with a standard vitamin mixture. This effect was abolished by supplementing dietary α-tocopherol at 500 I.U./kg diet. Membrane vit E was completely restored after 30 days (Table 1), whereas serum levels, despite increasing 2.5 times, remained significantly lower (p < 0.01) than those in CO-fed control mice. This effect of dietary FO may be associated with the fact that serum tocopherol levels are closely correlated with lipidemia (Farrel and Bieri, 1975) and dietary FO lowers blood lipids (Band and Dyerberg, 1980; Mouri et al., 1984; Phillipson et al., 1985). Hence, the large decline in vit E levels without expressing any outward symptoms may reflect a greater need for antioxidants in FO-fed animals. Vit E requirements were shown to be related to the tissue PUFA content (Witting and Horwitt, 1964; Dillard et al., 1978), and various fish oils elevate markedly the proportion of highly unsaturated fatty acids in subcellular membrane of liver (Table 2; Hammer and Wills, 1978), heart and brain (Tahin et al., 1981) and in macrophages (Leslie et al., 1985). However, our study clearly demonstrates that an increased requirement was not reflected by the double-bond index or the ratio of arachidonic/linoleic which both remained equal to those found in CO feeding. We observed that Pc was a better indicator of the membrane PUFA content as it was approximately doubled in FO-fed mouse liver microsome fractions. Our data obtained from *in vitro* peroxidation paralleled the calculated Pc's. The low Pc's of CO-fed mouse membrane fractions were accompanied by low rates of LPO and the lag period of induction previously reported in rat (Burk, 1985) was also observed in B/W mice. Feeding fish oil increased Pc and LPO but membrane fraction peroxidation was inversely correlated with its vit E content, as shown in Figs. 1 and 2.

A major finding of this study is that vit E is a potent inhibitor of LPO in FO-fed animal membranes exposed to 2 different pro-oxidant stimuli. However, in vit E repleted, FO-fed mouse membrane fractions, lipid peroxide accumulation was more effectively prevented under the non-enzymatic conditions (Fig. 1) as free-radical protection exerted by some antioxidants is known to vary with pro-oxidant conditions (Willson, 1983). Time course studies of LPO showed that the antioxidant ability of vit E had a longer duration in the presence of Fe/ascorbate than in the presence of NADPH-Fe-ADP, suggesting that tocopherols inactivated under the action of peroxy and alkoxy radical oxidation (Lambelet and Loliger, 1984) were regenerated reductively by the ascorbate (Liebler et al., 1986). Hammer and Wills (1978) also observed decreased microsomal lipid peroxidation in vit E supplemented herring oil-fed rats but vit E levels were not measured. In the study of Mouri et al., (1984) vit E did not reduce the liver TBA-positive materials of FO-fed rats, an indication that 20 mg-tocopherol/kg diet was insufficient supplementation to reduce lipid peroxidation.

Comparison of LPO in membrane fractions from CO-fed and FO-fed, vit E repleted mice which contain nearly equal levels of vit E suggests that the rate and extent of LPO are chiefly influenced by fatty acid peroxidizability. However, in agreement with Cheeseman et al. (1986), this should be interpreted as a consequence of the lower ratio of vit E to highly polyunsaturated lipids in FO-fed animal membranes that may account for the decreased induction time of peroxidation in FO preparations (Figs 1 and 2).

In conclusion, we found that liver membrane fractions, and probably most cellular membrane fractions of FO-fed animals may remain more susceptible to LPO even when

TABLE 2. EFFECT OF DIETARY OIL AND α-TOCOPHEROL SUPPLEMENTATION UPON LIVER MEMBRANE FATTY ACID COMPOSITION (MOL %)

MAJOR FATTY ACIDS		MITOCHONDRIA			MICROSOME		
		CORN OIL[a]	FISH OIL[a]	FISH OIL[b]	CORN OIL[a]	FISH OIL[a]	FISH OIL[b]
16:0		18.5 ± 1.3*	24.54 ± 1.0	24.83 ± 0.9	19.90 ± 0.3*	24.11 ± 0.9	24.28 ± 0.3
16:1	9	0.97 ± 0.3	1.60 ± 0.03	1.56 ± 0.2	0.80 ± 0.1	1.71 ± 0.03	2.06 ± 0.1
18:0		16.15 ± 0.7	14.78 ± 0.5	16.81 ± 1.2	13.94 ± 1.4	15.18 ± 1.0	15.05 ± 0.9
18:1	9	7.98 ± 0.5	6.05 ± 0.14	6.15 ± 0.2	10.46 ± 1.6	6.72 ± 0.3	7.21 ± 0.7
18:2	6	20.11 ± 1.0**	8.04 ± 0.6	6.89 ± 0.5	23.14 ± 2.5**	7.21 ± 0.2	6.29 ± 0.3
18:3	3	0.57 ± 0.3	0.17 ± 0.1	0.39 ± 0.1	0.19 ± 0.1	0.32 ± 0.02	0.29 ± 0.07
20:4	6	15.82 ± 0.9**	5.55 ± 0.6	5.58 ± 0.02	13.82 ± 1.2	5.89 ± 0.4	5.43 ± 0.3
20:5	3	0.07 ± 0.1	7.77 ± 1.1	5.94 ± 0.8	0.22 ± 0.1	7.90 ± 1.24	5.89 ± 0.3
22:5	6	0.83 ± 0.1	trace	trace	0.71 ± 0.2	0.55 ± 0.03	0.29 ± 0.2
22:6	3	10.52 ± 1.6**	21.53 ± 0.7	23.12 ± 1.4	7.77 ± 0.6**	22.44 ± 1.02	23.28 ± 0.3
INDEX		5.18 ± 0.1	5.14 ± 0.2	4.86 ± 0.6	4.38 ± 0.3	5.59 ± 0.5	5.13 ± 0.2
20:4/18:2		0.79 ± 0.02	0.72 ± 0.05	0.85 ± 0.02	0.73 ± 0.02	0.82 ± 0.4	0.87 ± 0.09
Pc		176.7 ± 8.7**	250.3 ± 8.4	248.8 ± 9.1	130.3 ± 6.8**	260.8 ± 13.3	252.5 ± 3.8

Mean values are given ± S.E.M. for CO n=5 preparations and FO n=3-4 preparations. CO vs FO: * $p < 0.01$, ** $p < 0.001$.

[a] = 75 I.U./kg α-tocopherol acetate
[b] = 500 I.U./kg α-tocopherol acetate fed for 30 days prior to sacrifice

Index (double bond index) is the sum of the % of each unsaturated fatty acid multiplied by the number of double-bonds divided by the sum of the % of the saturated fatty acids.

Peroxidizability (Pc) = (% Monounsaturated x 0.025) + (% diunsat. x 1) x (% triunsat. x 2) + (tetraunsat. x 4) + (% pentaunsat. x 6) + (% hexaunsat. x 8)

adequate α-tocopherol supplementation maintains vit E concentrations. Without proper vit E supplementation, increased lipid peroxidation can promote free-radical-induced damage in long-term studies and possibly negate the therapeutic benefits of dietary fish oil. This is consistent with our earlier study in which a low vit E-supplemented FO diet lowered natural killer cell activity and increased the frequency of myocarditis lesions in mice injected with Coxsackie virus (Fernandes et al., 1987). Our studies strongly suggest that optimum vit E supplementation should be maintained in FO-feeding experiments.

ACKNOWLEDGEMENTS

We wish to thank Kathy C. Laro and Vikram Tomar for excellent technical assistance and Dr. Dean Troyer for his kind advice and help.

REFERENCES

Bang, H. O., and Dyerberg, J., 1980, Plasma lipids and lipoprotein pattern in Greenlandic west coast Eskimos, **Lancet**, i:1143.
Burk, R., 1985, Glutathione-dependent protection by rat liver microsomal protein against lipid peroxidation, **Biochim. Biophys. Acta.**, 757:21.
Cheeseman, K. H., Collins, M., Proudfoot, K., Slater, T. F., Burton, G. W., Webb, A. C., and Ingold, K. U., 1986, Studies on lipid peroxidation in normal and tumor tissues. The Novikoff rat liver tumor, **Biochem. J.**, 253:507.
Dillard, C. J., Litov, R. E., and Tappel, A. L., 1978, Effects of dietary vitamin E, selenium, and polyunsaturated fats on *in vivo* lipid peroxidation in the rat as measured by pentane production, **Lipids**, 13:396.
Farrel, P. M., and Bieri, J. G., 1975, Megavitamin E supplementation in man, **Am. J. Clin. Nutr.**, 28:1381.
Fernandes, G., 1987, Influence of nutrition on autoimmune disease, *in:* "**Aging and the Immune Response**," E. J. Goidl, ed., Marcel Dekker, Inc., New York.
Fernandes, G., Gaunt, C., Sandberg, L., Friedrichs, W. E., and Meydani, S. N., 1987, Effect of high fish oil intake on vitamin E levels, NK activity and susceptibility to viral-induced myocarditis in lupus-prone mice, **Arth. Rheum.**, 30:S123.
Folch, J., Lees, M., and Sloane-Stanley, G. H., 1957, A simple method for the isolation and purification of total lipids from animal tissues, **J. Biol. Chem.**, 226:497.
Hammer, C. T., and Wills, E. D., 1978, The role of lipid components of the diet in the regulation of the fatty acid composition of the rat liver endoplasmic reticulum and lipid peroxidation, **Biochem J.**, 174:585.
Holman, R. E, 1954, Progress in the chemistry of fats and other lipids, Vol. 2, R. T. Holman, W. O. Lundberg, and T. Malkin, eds., Academic Press, New York.
Kremer, J. M., Bigauoette, J., Michalek, A. V. et al., 1985, Effects of manipulation of dietary fatty acids on clinical manifestations of rheumatoid arthritis, **Lancet**, i:184.
Laganiere, S., and Yu, B. P., 1987, Anti-lipoperoxidation action of food restriction, **Biochem. Biophys. Res. Comm.**, 145:1185.
Lambelet, P., and Loliger, J., 1984, The fate of anti-oxidant radicals during lipid auto-oxidation II. The tocopheroxyl radicals, **Chem. Phys. Lipids**, 35:184.
Leslie, C. A., Gonnerman, W. A., Ullman, M. D., Hayes, K. C., Franzblau, C., and Cathcart, E. S., 1985, Dietary fish oil modulates macrophage fatty acids and decreases arthritis susceptibility in mice, **J. Exp. Med.**, 162:1336.
Liebler, D. C., Kling, D. S., and Reed, D. J., 1986, Antioxidant protection of phospholipid bilayers by α-tocopherol, **J. Biol. Chem.**, 261:12114.
Mouri, K., Ikesu, H., Esaka, T., and Isarashi, O., 1984, The influence of marine oil intake upon levels of lipids, α-tocopherol, and lipid peroxidation in serum and liver of rats, **J. Nutr. Sci. Vitaminol.**, 30:307.
Ohkawa, H., Ohishi, N., and Yagi, K., 1979, Assay for lipid peroxides in animal tissues by thiobarbituric acid reaction, **Anal. Biochem.**, 95:351.
Phillipson, B. E., Rothrock, D. W., Connor, W. E., Harris, W. S., and Illingworth, D. R., 1985, Reduction of plasma lipids lipoproteins and apoproteins by dietary fish oils in patients with hypertriglyceridemia, **N. Eng. J. Med.**, 312:1210.

Prickett, J. D., Robinson, D. R., and Steinberg, A. D., 1981, Dietary enrichment with the polyunsaturated fatty acid eicosapentaenoic acid prevents proteinuria and prolongs survival in NZ₁ x NZWF₁ mice, **J. Clin. Invest.**, 68:556.

Siess, W., Scherer, B., Bohlig, B., Roth, P., Kurzmann, I., and Weber, P. C., 1980, Platelet-membrane fatty acids, platelet aggregation, and thromboxane formation during a mackerel diet, **Lancet**, i:441.

Stubbs, C. D., and Smith, A. C., 1984, The modification of mammalian membrane polyunsaturated fatty acid composition in relation to membrane fluidity and function, **Biochim. Biophys. Acta**, 779:89.

Tahin, Q. S., Blum, I., and Carafoli, E., 1981, The fatty acid composition of subcellular membranes of rat liver, heart and brain. Diet-induced modifications, **Eur. J. Biochem.**, 121:5.

Taylor, S. L., Lambden, M. P., and Tappel, A. L., 1976, Sensitive fluorometric method for tissue tocopherol analysis, **Lipids,** 11:530.

Wilson, R. L., 1983, Free radical protection: why vitamin E not Vitamin C, beta-carotene or glutathione?, **CIBA Found. Symp.**, 101:19.

Witting, L. A., and Horwitt, M. K., 1964, Effect of degree of fatty acid unsaturation in tocopherol deficiency-induced creatinuria, **J. Nutr.**, 82:19.

IMMUNITY AND DISEASE RESISTANCE IN FARM ANIMALS FED VITAMIN E

SUPPLEMENT

Robert P. Tengerdy

Department of Microbiology
Colorado State University
Fort Collins, CO 80523

INTRODUCTION

The optimal physiologic functioning of an animal organism requires a restricted but well balanced diet, including protein-calorie nutrients, vitamins and minerals. Vitamin E, in conjunction with selenium, as an antioxidant and cell membrane stabilizing agent affects many physiological functions, including the defense mechanism of the body. Vitamin E deficiency leads to impaired immune functions and decreased disease resistance. Vitamin E supplementation, on the other hand, may lead to improved immunity and increased disease resistance.

The purpose of this paper is to scrutinize data from our own laboratory and others on the effect of vitamin E supplementation on immune responses and disease resistance, especially in farm animals used in modern production practice. Earlier reviews on this subject concentrated on available data mostly from laboratory animals (Tengerdy, 1980; Tengerdy et al., 1981; Tengerdy et al., 1984; Tengerdy, 1986), thus a review oriented toward farm animals is both timely and may have a practical use.

ESTABLISHING THE OPTIMAL LEVEL OF VITAMIN E FOR IMMUNE RESPONSE AND DISEASE RESISTANCE

In comparing data in the literature about the effect of vitamin E on immunity and disease resistance, there is some confusion and contradiction in interpreting the data, because different workers use different doses and different regimes of administration. Although vitamin E supplementation levels have been recommended for a number of species, as shown in Table 1, the recommended dose is based on the level that avoids deficiency symptoms, rather than one that gives optimal performance. The actual optimal dose of vitamin E for maximum immune enhancement and disease protection, in our experience, is 3-6 times higher than the recommendations shown in Table 1.

The problem is aggravated in modern farm animal production, where the demand for maximum meat, egg or milk production by genetically improved animals requires overfeeding at 2-8 times the maintenance requirement. This alters the vitamin and mineral requirements in the diet. An example is shown in Table 2, how improved feed utilization may alter vitamin E requirement in the feed.

Other factors that may increase vitamin E requirement are PUFA, nitrites, ionophores, mycotoxins, high concentrate rations, all having daily occurrence in modern farm animal nutrition. For example, every gram increase of PUFA in the diet requires 1-3 mg vitamin E to compensate. Increased production of meat, eggs and milk proportionally increases the demand for vitamin E, as illustrated in Table 3. Stress by transportation, heat, noise, crowding in pens and feedlots also contribute to an increased vitamin E requirement. In our experiments with chicken, mice and sheep we consistently found an optimal dose of 300 mg/kg diet, although a

Table 1. Recommended levels of vitamin E supplementation.

Poultry	20-60 mg/kg diet
Cattle	15-60 mg/kg diet
Fish	70-80 mg/kg diet
Racehorse	150 mg/kg diet

Assuming that average natural ration contributes at least 10 mg/kg.

Source: Hoffman La Roche Technical Information.

Table 2. Feed consumption and vitamin E requirement.

		1970		1980	
Species	Requirement mg	Feed consumption	Dietary level mg/kg	Feed consumption	Dietary level mg/kg
Pig	50	2.72	18	2.0 kg	25
Laying hen	2.75	125 g	22	110 g	25
Cattle	300	15 kg	20	12 kg	25

Source: Hoffmann La Roche Technical Information

Table 3. Increased demand for vitamin E by increased animal production.

Product	Production 1960	(hd,yr) 1980	Percent increase	Vitamin E 1960	(hd/day) 1980
Milk	3000 l	4000 l	25	300 mg	375 mg
Egg	180	250	30	2.75 mg	3.85 mg

Source: Hoffmann La Roche Technical Information

biphasic response was always evident, that is 150 mg/kg also gave acceptable results (Tengerdy et al., 1984).

In farm animal production practice the cost/benefit ratio has to be considered for an economically practical yet still effective dose of vitamin E. In poultry, reasonably good results were obtained in our laboratory with 100 mg/kg diet, Morrill and Reddy obtained fair protection with 125 mg/hd, day over a 24 week period (Morrill and Reddy, 1987). The length of treatment varies: for vaccination it is recommended to feed for 3-4 weeks prior to vaccination, for maintenance in a high stress or high natural infection situation the treatment should last while the cause persists.

Since the proper feeding level of vitamin E depends on so many factors, it is recommended that the plasma vitamin E level and Se level be regarded as the measure of adequacy of a given animal. The correlation between plasma levels and physiologic performance are generally better than between feed levels and performance. Vitamin E supplementation should always be done in the presence of an adequate Se level.

Representative examples of the use of vitamin E in farm animals for increasing disease protection are given in Table 4. The doses applied and the duration and mode of administration vary greatly, so do the actual results.

Table 4. Vitamin E response to infections in farm animals.

Species	Administration	Disease	Effect	Ref
Calves	125 mg/day, 24 weeks orally + Se	Bovine herpes virus	Decreased virus titer high LSI[1] counteracting stress	9
Calves	1 g/day, 6 weeks orally + Se		High IgG, IgA, low IgM, high LSI	18
Jersey calves	0.5-2.5 g/day 30 days	Oral necrobaccillosis diarrhea P. multocida	better recovery increased titer to vaccine	23
Lactating cows	1 g/day + 3 mgSe/day in dry period 0.5 g/day + 6 mg SE/day during lactation	mastitis	30-50% reduction of clinical mastitis, high plasma E, Se, glutathione peroxidase; increased PMN function	8
Pigs	20-100 mg/kg diet	E. coli	increased resistance	19
Pigs	200 mg/hd, day	dysentery	increased resistance, increased CMI[2]	20 21
Sheep	300 mg/kg diet 3 weeks	enterotoxemia	increased resistance, increased antibody to C. perfringens	12
Sheep	300 mg/kg diet 3 weeks	Chlamydia	increased resistance	22

[1]LSI-lymphocyte stimulation index
[2]CMI=cell mediated immunity

It appears that daily administration of relatively lower doses mixed with the diet are better than large doses administered at once or at intervals orally or by injection.

The responses reported include enhanced humoral immunity, manifested in elevated IgG and IgA levels and a concomitant shift from IgM to IgG; enhanced phagocytosis by polymorphonuclear cells (PMN, mostly neutrophils); enhanced cell mediated immunity, manifested by increased stimulation of lymphocytes to mitogens and antigens. The results in farm animals corroborate earlier findings in laboratory animals about the mechanism of disease protection by vitamin E (Tengerdy et al., 1984; Corwin and Gordon, 1982). Vitamin E primarily enhances the activity of T helper cells and macrophages through stabilizing the lipids in the cell membrane, and modulating cell receptor functions through regulating prostaglandin, thromboxane and leukotriene biosynthesis (Tengerdy et al., 1984; Likoff et al., 1981; Lawrence et al., 1985). Vitamin E also activates the bactericidal oxidative mechanism of PMN (Boxer, 1986). Vitamin E is most effective in enhancing immune phagocytosis by PMN, therefore, its use is most beneficial in infectious diseases, mostly bacterial, where immune phagocytosis is the chief defensive mechanism. In bacterial diseases, where one or more elements of this mechanism are blocked, vitamin E is ineffective. For instance, in mastitis caused by *Corynebacterium bovis*, vitamin E is ineffective, probably because the bacteria are localized in the streak canal of the teat, inaccessible to PMN (Smith and Conrad, 1987). The beneficial effect of vitamin E in infectious diseases involving primarily cell mediated immunity has not been demonstrated convincingly yet, although vitamin E definitely enhances the activity of lymphocytes and macrophages (Corwin and Gordon, 1982).

Probably the most beneficial effect of vitamin E is manifested under stress. In farm animal practice this is an important consideration, because modern animal management involves unavoidable stress during shipping, crowding in feedlots or cages, exposure to high temperature, noise, rough handling.

Table 5. Correlation between prostaglandin (PG) levels and mortality in *E. coli* infected chickens.

Treatment[1]	PG in bursa, ug/g wet tissue[2]			Percent[3] mortality
	PGF$_2$	PGE$_2$	PGE$_1$	
Control	71.9±8	22.5±1.9	94.0±14.8	80
Vitamin E	36.2±3.8	21.2±2.7	32.5±6.4	36
Aspirin	39.7±4.6	12.2±1.0	70.0±5.9	42
Vitamin E + Aspirin	12.4±3.6	2.9±1.0	23.3±7.7	0
Non-infected	44.2	27.4	46.6	0

[1] All chickens except noninfected controls were injected with 1×10^9 *E. coli* . Vitamin E dietary supplement 300 mg/kg diet; aspirin 50 mg/kg body weight, injected intraperitoneally on the day before infection and dailty thereafter.
[2] n = 8. PG radioimmunoassays from bursa taken from sacrificed birds 5 min after infection.
[3] Counted 2 days after infection.

Source: Likoff et al. 1981

Vitamin E appears to counteract certain effects of stress, such as the immunosuppressive effect of elevated cortisol levels and associated increase in prostaglandins (Watson and Petro, 1982). We reported earlier the correlation between depressed prostaglandin levels and immunoenhancement (Table 5) (Likoff et al., 1981; Lawrence et al., 1985). Recent data show that vitamin E, prostaglandins and dietary fat interact in immunomodulation (Table 6). It is interesting to note that a saturated fat depressed PGE$_2$ and elevated antibody titer more than a PUFA.

In assessing the expected benefit of vitamin E supplementation for disease protection, other nutritional interactions beside the mentioned fat interaction should be considered. Nutritional interactions may change the effective vitamin E requirement. Not only PUFA increases the vitamin E requirement, but the high concentrate rations used in cattle fattening in feedlots also require extra compensation by vitamin E (Stuart, 1987). The antagonistic interaction of high levels of vitamin E and A was reported earlier (Tengerdy and Brown, 1977). Bendich has reported a positive interaction. between vitamin E and C (Bendich, 1987), and unconfirmed report suggest a possible positive E and C interaction in trout (R. Anderson, personal communication). An adequate level of Se is indispensable for potentiating vitamin E as discussed elsewhere in this symposium.

Table 6. Interaction of dietary lipids and vitamin E on prostaglandin (PGE$_2$) level and antibody titer in cockerels.

Diet	PGE$_2$ (pg/mg) in bursa	Antibody Log$_2$ titer
control	33.1	2.7
control + E	21.0	2.9
3% safflower oil	32.6	2.7
3% safflower oil + E	22.6	2.3
3% beef tallow	30.3	2.8
3% beef tallow + E	15.5	3.0

Source: Nockels, C.F. unpublished data.

ALTERNATE USE OF VITAMIN E FOR IMMUNOMODULATION AND DISEASE PROTECTION

It appears that the required high dietary supplementary dose of vitamin E for effective disease protection may not be affordable to many animal producers, especially in less developed countries. An alternate and more cost effective use for vitamin E administration was found in our laboratory in the past few years, the use of vitamin E in water-in-oil adjuvant vaccines for immunomodulation and increased disease protection (Tengerdy et al., 1983; Afzal et al., 1984; Afzal et al., 1986). In such adjuvants, the antigen is released slowly and gradually from the oily depot, and the vitamin E locally stimulates macrophages and T helper cells at the site of injection (Afzal et al., 1984). A single dose of 850 mg vitamin E in such an adjuvant preparation gave better protection against enterotoxemia of sheep than the dietary administration of 300 mg/kg fed for 3 weeks (Table 7).

The vitamin E adjuvant was tested successfully in Colorado against ram epididymitis caused by *Brucella ovis* (Table 8) and is being tested currently in a large field trial in Peru (Figure 1).

Table 7. Comparison of *Clostridium perfringens* antitoxin D titers in vitamin E adjuvant and conventionally immunized lambs.

Group	Mean antibody titer[1]	
	before vaccination	after vaccination (7 days)
Vaccinated, control diet	0.312±09(n=12)	0.513±0.22(n=12)
Vaccinated, vitamin E diet	0.332±0.10(n=12)	0.725±0.14(n=12)
Vitamin E adjuvant, control diet	0.446±0.14(n=3)	1.64±0.25(n=3)
Vitamin E adjuvant, Vitamin E diet	0.354±0.12(n=3)	1.046±0.24(n=3)\

[1]Measured in an ELISA test. Titer expressed in absorbance units at 405 nm and 1:200 serum dilution. The pelleted diet was supplemented with 300 mg/kg vitamin E and fed for 5 weeks.

Source: Tengerdy et al., 1983

Table 8. Correlation of humoral immunity and infection in rams vaccinated with *B. ovis* against epididymitis.

Group	No. in Group	No. Infected	Percent Overall Infectivity	Peak ELISA[1] Titer ± S.D.
Bacterin[2]	9	4	44.4	0.27±0.15
B. ovis-Vit E	9	2	22.2	0.42±0.25
B. ovis-FIA[3]	9	4	44.4	0.35±0.16
Vit. E placebo	8	3	37.5	0.14±0.08
FIA placebo	8	5	62.5	0.15±0.09
Aluminum hydroxide control	9	6	66.7	0.26±0.18

[1]Enzyme linked immunosorbent assay: titer is expressed in O.D. units at 488 nm and 1:200 serum dilution.
[2]Killed *B. ovis* cells in Al(OH)$_3$.
[3]Freund's incomplete adjuvant.

Source: Afzal et al., 1984

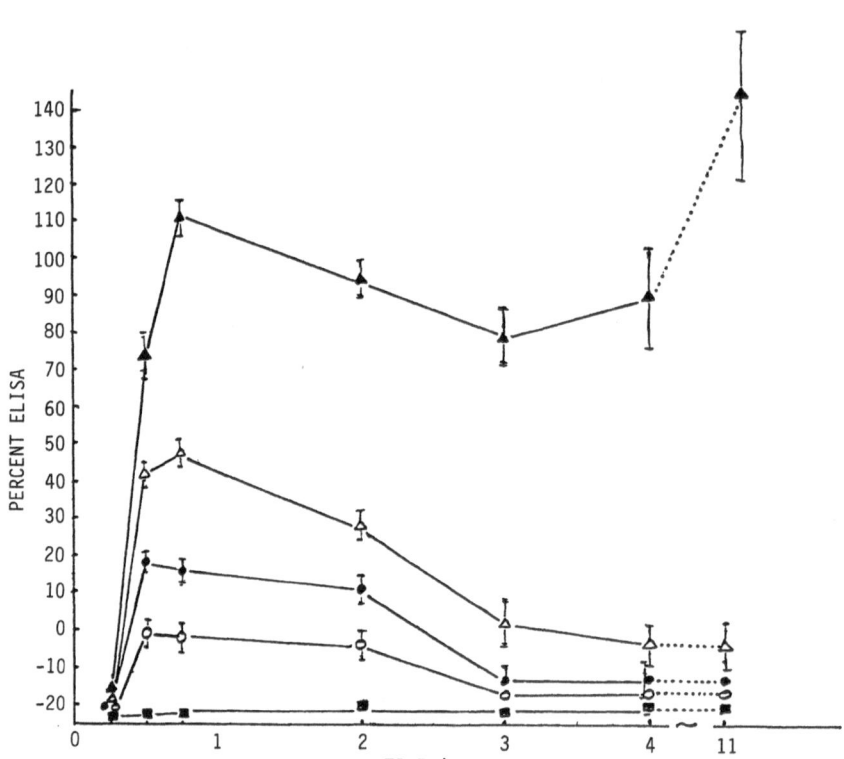

Figure 1. ELISA antibody titers of rams immunized with *Brucella* vaccines. *B. ovis* vitamin E adjuvant vaccine; Al (OH)$_3$ precipitated *B. ovis* bacterin; Rev 1 (*B. melitensis*) in vitamin E adjuvant; Rev 1; Control; n = 10:20. Titers are expressed as optical density in percentage of reference standards.

CONCLUSIONS

A survey of the literature and our own research indicate that dietary supplementation of vitamin E may be beneficial to increase disease resistance in farm animals, cattle, pigs, poultry, and sheep, especially under stress conditions. The minimal effective dose has to be established for each case, taking in consideration feeding regimes and management conditions, using plasma vitamin E levels as guidance criterion. An alternate use of vitamin E in adjuvant vaccines is very effective.

REFERENCES

Afzal, M., Hussain, M., Khan, K.M.N., and Munir, A., 1988, Effect of vitamin E and selenium on immunity in newborn Jersey and buffalo calves, **Pakistan Vet. J.,** (in press).

Afzal, M., Tengerdy, R. P., Brodie, S. J., DeMartini, J. C., Ellis, R. P., Jones, R. L., and Kimberling, C. V., 1986, The immune response in rams experimentally infected with *Brucella ovis*, **Res. Vet. Sci.,** 41:85.

Afzal, M., Tengerdy, R. P., Ellis, R. P., Kimberling, C. V., and Morris, C. J., 1984, Protection of rams against epididymitis by a *B. ovis*-vitamin E adjuvant vaccine, **Vet. Immunol. Immunopath.,** 7:293.

Bendich, A., 1987, Role of antioxidant vitamins on immune function, Roche Tech. Symposium, March 11, Daytona, Florida.

Boxer, L. A., 1986, Regulation of phagocyte function by alpha-tocopherol, **Proc. Nutr. Soc.,** 45:333.

Cipriano, J. E., Morrill, J. L., and Anderson, N. V., 1982, Effect of dietary vitamin E on immune responses of calves, **J. Dairy Sci.,** 65:2357.

Corwin, L. M., and Gordon, R. V., 1982, Vitamin E and immune regulation, **N.Y. Acad. Sci.,** 393:437.

Ellis, R. P., and Vorties, M. W, 1976, Effect of supplemental dietary vitamin E on the serologic response of swine to *E. coli* bacteria, **J. Am. Vet. Med. Assn.,** 168:231.

Heinzerlilng, R. H., Tengerdy, R. P., Wick, L. L., and Lueker, D. C., 1974, Vitamin E protects mice against *Diplococcus pneumoniae* type I infection, **Infection & Immunity,** 10:1292.

Larsen, H. J., and Tollersrud, S., 1981, Effect of dietary vitamin E and selenium on the phytohemaglutinin responses of pig lymphocytes, **Res. Vet. Sci.,** 31:301.

Lawrence, L. M., Mathias, M. M., Nockels, C. F., and Tengerdy, R. P., 1985, The effect of vitamin E on prostaglandin levels in the immune organs of chicks during the course of an *E. coli* infection, **Nutr. Res.,** 5:497.

Likoff, R. O., Guptill, D. R., Lawrence, L. M., McKay, C. C., Mathias, M. M., Nockels, C. F., and Tengerdy, R. P., 1981, Vitamin E and aspirin depresses prostaglandins in protection of chickens against *E. coli* infection, **Am. J. Clin. Nutr.,** 34:245.

Morrill, J. L, and Reddy, P. G., 1987, Effect of vitamin E on immune responses and performance of dairy calves, Roche Techn. Symposium, March 11, Daytona, Florida.

Nockels, C. F., 1974, Protective effects of supplemental vitamin E against infection, **Fed. Proc.,** 38:2134.

Smith, L. K., and Conrad, H. R., 1987, Vitamin E and selenium supplementation for dairy cows, Roche Techn. Symposium, March 11, Daytona, Florida.

Stuart, R. L., 1987, Factors affecting vitamin E status of beef cattle, Roche Techn. Symposium, March 11, Daytona, Florida.

Teige, J., Tollersrud, S., Lund, A. and Larsen, H. J., 1982, Swine dysentery: the influence of dietary vitamin E and selenium on the clinical and pathological effects of *Treponema hyodysenteriae* infection in pigs, **Res. Vet. Sci.,** 32:95.

Tengerdy, R. P., 1980, Effect of vitamin E on immune responses, *in*: **"Vitamin E, a Comprehensive Treatise,"** L. J. Machlin, ed., Marcel Dekker, Inc., New York.

Tengerdy, R. P., 1986, Nutrition, immunity and disease resistance, *in*: **Proceedings, Sixth International Conference on Production Disease in Farm Animals,** Belfast, N. Ireland.

Tengerdy, R. P., and Brown, J. C., 1977, Effect of vitamin E and A on humoral immunity and phagocytosis in *E. coli* infected chicken, **Poultry Sci.,** 56:957.

Tengerdy, R. P., Mathias, M. M., and Nockels, C. F., 1981, Vitamin E, immunity and disease resistance, *in*: **"Diet and Resistance to Disease,"** M. Phillips and A. Baetz, eds., Plenum Press, New York.

Tengerdy, R. P., Mathias, M. M., and Nockels, C. F., 1984, Effect of vitamin E on immunity and disease resistance, *in*: **"Vitamins, Nutrition and Cancer,"** K. N. Prasad, ed., Karger, Basel.

Tengerdy, R. P., Meyer, D. L., Lauerman, L. H., Lueker, D. C., and Nockels, C. I., 1983, Vitamin E enhances humoral antibody response to *Clostridium perfringens*, type D, in sheep, **Brit. Vet. J.,** 139:147.

Watson, R. R., and Petro, T. M., 1982, Cellular immune response, corticosteroid levels and resistance to *Listeria monocytogenes* and murine leukemia in mice fed a high vitamin E diet, **Ann. N.Y. Acad. Sci.,** 393:205.

POSSIBLE ROLES FOR ZINC IN DESTRUCTION OF *TRYPANOSOMA CRUZI* BY TOXIC OXYGEN METABOLITES PRODUCED BY MONONUCLEAR PHAGOCYTES

J. M. Cook-Mills, J. J. Wirth, and P. J. Fraker

Department of Biochemistry
Michigan State University
East Lansing, MI 48824

SUMMARY

The effects of a single nutrient deficiency on immune function is now most extensively characterized using the dietary zinc deficient murine model. Deficiencies in zinc have rapid adverse effects on host defenses of humans and rodents. This impaired defense seems to be, in part, the result of a reduction in number of lymphocytes available for surveillance since residual lymphocytes are able to carry out many normal functions. *In vitro*, the lymphocytes were able to proliferate at a normal rate as well as produce antibodies or interleukin 2 in response to mitogens or antigens even when cultured in autologous serum to reduce the possibility of restoration of zinc deficient functions. Conversely, mononuclear phagocytes (MNP) from deficient mice had a significantly reduced capacity to associate with and kill the parasite *Trypanosoma cruzi* (*T. cruzi*) which causes Chaga's disease. Moreover, indicating the specificity of the deficient function, a short incubation of $ZnCl_2$ but not other metals completely restored the capacity of MNP from deficient mice to take up and kill *T. cruzi*. Dependency on H_2O_2 production by the MNP's oxygen burst for killing of *T. cruzi*.suggested that MNP from zinc deficient mice might produce smaller amounts of H_2O_2. The possibility that zinc might play an integral role in the oxygen burst seemed evident from the ability of zinc to quickly restore the killing capacity of MNP from the zinc deficient mice. Further, the renewed interest in the role of metals in the production of highly reactive oxidants in biological systems prompted a literature search to identify enzymes and/or reactions known to be involved in the generation of oxygen radicals or toxic oxygen metabolites that might be zinc dependent. The literature review provided herein indicates many possible roles for zinc in the generation of toxic oxygen species. The data indicated that normal levels of H_2O_2 are produced by MNP from zinc deficient mice. The amount of H_2O_2/mg macrophage protein is normal in response to phorbol or opsonized zymosan but reduced in response to direct stimulation by *T. cruzi*. However, the reduced H_2O_2 production by *T. cruzi*-stimulated zinc deficient MNP was due to reduced stimulation as a result of fewer *T. cruzi* associated with the MNP. Thus, H_2O_2 levels/parasite were the same as zinc adequate controls. Yet, this does not preclude the possibility that reduced killing of *T. cruzi* by MNP from zinc deficient mice may be due to a function for zinc in the actual killing process or in the production of some other agent important in the killing of *T. cruzi*.

Zinc Deficiency

Deficiencies in zinc are frequently observed in the human population in Third World nations as well as in the USA (Prasad, 1984; Sandstead, 1973). Consumption of diets high in certain grains, cereals or refined foods can create a marginal dietary deficiency in zinc, especially if consumption of meat and animal products is low (Prasad, 1984; Sandstead, 1973). A daily consumption of zinc is required since there are no significant bodily stores for this metal (Prasad, 1979). Deficiencies in zinc can accompany many disease states such as gastrointestinal disorders, renal disease, alcoholism, liver disease, sickle cell anemia, etc. (Prasad, 1984; Prasad,

111

1979). The many biochemical roles for zinc include cofactor activity for over 100 metalloenzymes such as RNA and DNA polymerase (Vallee and Galdes, 1984), binding to nucleic acid binding proteins which partake in gene expression (Berg, 1986), and perhaps participation in membrane integrity and protein synthesis (Prasad, 1979). Potential biochemical roles for zinc in the oxygen burst by mononuclear phagocytic cells (MNP) will be discussed later. The prevalence of the deficiency in the human population has fostered a variety of studies using primarily rodents as the model to better define the effects of the deficiency on growth, development and various cellular and tissue functions. Our studies have focused on the effects of zinc deficiency on immune function.

Effect of Zinc Deficiency on Immune Function: A Brief Review

Zinc deficiency has a rapid and adverse effect on immune function both in rodents and humankind (Fraker et al., 1986; Gershwin et al., 1979; Fernandes et al., 1979; Fraker et al., 1977; Good, 1981). Using the mouse as a model, because of its close immunological relationship to mankind, extensive information on the impact of a moderate period of suboptimal intake of zinc on host defense systems has been generated (Fraker et al., 1986; Gershwin et al., 1979; Fernandes et al., 1979; Fraker et al., 1977). Briefly, six week old A/J female mice are fed *ad libitum* a biotin fortified egg white diet containing either deficient (0.8ug Zn/g) or adequate (27ug Zn/g) levels of zinc. Since inanition accompanies zinc deficiency, a third group are restricted in consumption of zinc adequate diet to that consumed by zinc deficient mice. At the time of analysis of immune statis, 30 days, the deficient group is further subdivided into two groups; mice at 67% body weight of controls are assigned to the severely zinc deficient group and mice at 75% body weight of controls are moderately zinc deficient. A thirty-day period of suboptimal intake of zinc by the young adult mouse reduces antibody mediated responses to either T-cell dependent or B-cell independent antigens by 40 to 60 percent (Fraker et al., 1986; Fernandes et al., 1979; Fraker et al., 1977) and reduces cell-mediated responses in tumor defense or delayed type hypersensitivity reactions (Fernandes et al., 1979; Fraker et al., 1982). Also, marked atrophy of the thymus and lymph nodes (Fraker et al., 1978) is accompanied by a 30-50% reduction in the total numbers of lymphocytes and MNP (Fraker et al., 1978).

The substantial reduction in the number of leukocytes brought to fore the question of whether the residual cells of the immune system of the zinc deficient mouse were fully functional especially since many enzymes of the cell are known to be zinc dependent. Studies of lymphocytes cultured in autologous serum to reduce the opportunity for restoration of zinc dependent function indicates that the residual lymphocytes of the deficient mice are fully functional (Fraker et al., 1988; Fraker et al., 1987). Proliferation and interleukin 2 production by mitogen or allogeneic cell-stimulated splenic T-cells from deficient mice was the same or better than splenic T-cells from adequately fed mice regardless of whether the zinc level of the culture system was adequate or limiting (Fraker et al., 1988; Fraker et al., 1987). Likewise, β-cells from deficient mice proliferated normally and produced normal levels of antibody when challenged with mitogens or actual antigens (Fraker et al., 1988). Thus, since the residual cells were capable of performing a variety of functions, it suggested that a primary effect of zinc deficiency on immunity was to reduce defense capacity via the reduction in the total numbers of lymphocytes available for service against an immunogenic challenge.

Initial tests of macrophage status of the deficient mice indicated that there were significant changes in Fc and complement receptor expression of peripheral blood and splenic mononuclear phagocyte (MNP) but not of peritoneal MNP (Wirth et al., 1984). Tests of the ability of the peripheral blood and peritoneal MNP to phagocytize polystyrene beads revealed no difference among the dietary groups. However, these results were suspect in the light of other experiments to be discussed. In an effort to show the degree to which zinc deficiency could impair host defense, mice from the dietary groups challenged on day 10 of the diet regimen with a sublethal dose of the parasite *Trypanosoma cruzi (T. cruzi)*. This particular parasite causes Chaga's disease in millions of South Americans each year. Within twelve days of infection, eighty percent of the zinc deficient mice had died exhibiting twenty times greater numbers of *T. cruzi* in their blood than the zinc adequate or restricted-fed mice (Fraker et al., 1982). At 30 days post-infection 85% of the infected zinc deficient mice were dead. No deaths had occurred in the uninfected zinc deficient mice or the infected adequately fed mice while the infected restricted fed mice had experienced 10% mortalities. This experiment dramatically demonstrated the increased vulnerability of the nutritionally deficient mice to pathogenic infections.

Since it is also known that MNP are a vital first line of defense against *T. cruzi*, the inability of the deficient mice to mount even a minimal defense against the parasite prompted us to look again at MNP function. MNP from deficient, adequate or restricted fed mice were pretreated for 1 hour with ample (10 µg $ZnCl_2$/ml) or limiting quantities of zinc (<0.3 µg Zn/ml) before infection with *T. cruzi*. The results of using a natural pathogen were markedly contrasted to the data obtained with polystyrene beads. As can be seen in Table 1 Experiment 1, the degree of association of *T. cruzi* with MNP from moderately deficient mice was moderately reduced from controls. MNP from severely deficient mice was reduced by greater than 60%. Killing of *T. cruzi* by MNP at six hours post-infection were even more interesting. By then, MNP from zinc adequate and restricted fed mice had killed 40-50% of the associated parasites. MNP from

Table 1. Effect of Pretreatment with Zinc Chloride on the Ability of Peritoneal Macrophages from Various Dietary Treatment Groups to Associate with and Kill *T. cruzi*

Dietary Group	1 hour Pretreat w/$ZnCl_2$ (µg/ml)	Experiment 1		Experiment 2	
		# *T. cruzi* Associated w/ MNP[a]	% *T. cruzi* Killed by MNP	# *T. cruzi* Associated w/ MNP[a]	% *T. cruzi* Killed by MNP[c]
Severe	0.0	34±4[d,e]	+ 31	40±5	0
Moderate	0.0	79±2[e]	- 18	48±8	0
Restricted	0.0	104±6	- 46	89±15	-61
Control	0.0	100±6	- 46	100±15	-66
Severe	1.0	nd[f]	nd	47±5	-51
Moderate	1.0	nd	nd	49±7	-15
Restricted	1.0	nd	nd	82±8	-58
Control	1.0	nd	nd	100±8	-57
Severe	2.5	nd	nd	70±3	-28
Moderate	2.5	nd	nd	66±20	-34
Restricted	2.5	nd	nd	100±17	-39
Control	2.5	nd	nd	100±11	-39
Severe	5.0	nd	nd	90±8	-54
Moderate	5.0	nd	nd	79±11	-54
Restricted	5.0	nd	nd	86±4	-68
Control	5.0	nd	nd	100±16	-69
Severe	10.0	83±8	- 59	nd	nd
Moderate	10.0	90±4	- 46	nd	nd
Restricted	10.0	80±2	- 36	nd	nd
Control	10.0	100±9	- 47	nd	nd

[a]Percent of control for number of *T. cruzi*/100 NMP at 0 hours.
[b]Percent change in number of *T. cruzi*/100 NMP from 0 to 6 hours.
[c]Percent change in number of *T. cruzi*/100 MNP from 0 to 24 hours.
[d]Values are means ± SD of six to eight mice.
[e]The difference between this value and the value of the control is statistically significant at p<0.05 or greater.
[f] nd indicates not determined.

moderately deficient mice had killed only 18% of associated *T. cruzi* suggesting the parasites were able to proliferate while engulfed in the MNP. The results indicated that the zinc deficiency may have significantly impaired the oxygen burst of the MNP which is essential to the killing of the parasite. More amazingly, preincubation of MNP from the deficient mice for one hour in the presence of zinc chloride at about five times physiological levels (10 µg $ZnCl_2$/ml) completely restored all of these functions (Table 1 Experiment 1). Preincubation with 1, 2.5, or 5 µg $ZnCl_2$/ml moderately restored association and killing of *T. cruzi* by MNP from zinc deficient mice (Table 1 Experiment 2). The number of associated *T. cruzi* per 100 MNP was the same for MNP of deficient mice and controls after exposure to zinc. Perhaps more importantly, preincubation with zinc completely restored the capacity of MNP from deficient mice to kill *T. cruzi* (Table 1). Other metals such as manganese, nickle, and copper were unable to restore these functions (data not shown).

It is evident from these experiments that the MNP from zinc deficient mice had functional processes that were impaired due to the limiting availability of zinc in the host environment. It was intriguing that only a short time of incubation with $ZnCl_2$ was required to restore both association of and capacity to kill the parasite by MNP from deficient mice. As will be discussed, the destruction of *T. cruzi* has been thought to be heavily dependent on the oxygen burst, particularly the production of H_2O_2. In addition, there is increasing evidence that metals may play a greater role than previously thought in the production of toxic oxygen metabolites by cells and tissues. Thus, it occurred to us that zinc itself may play an integral role in the respiratory burst of the MNP.

Production of Toxic Oxygen Metabolites by MNP

Contact with a variety of foreign substances, particularly pathogens, initiates the oxygen burst in MNP. The burst is a critical set of reactions which generate toxic oxygen metabolites responsible for the killing of most pathogens (Figure 1). The chemical phorbol myristate acetate (PMA) and opsonized zymosan, an antibody coated yeast cell extract, are frequently used by investigators to activate MNP for the purpose of initiating the oxygen burst. Thus, this discussion will necessarily revolve around the available experiments done with artificial probes.

The oxygen burst as it is currently understood is depicted in Figure 1. Data indicate the PMA and opsonized zymosan promote activation of NADPH oxidase, an enzyme located in the plasma membrane, which catalyzes the production of superoxide (O_2-) from molecular oxygen. The O_2-, then, either nonenzymatically dismutates to H_2O_2 or superoxide dismutase catalyzes the dismutation. O_2- and H_2O_2 are scavenged by superoxide dismutase and catalase, respectively, to protect the macrophage. Glutathione peroxidase also scavenges H_2O_2 while oxidizing glutathione. Within the phagosome/phagolysosome, H_2O_2 reacts with O_2- to produce O_2, OH- and ·OH. The O_2-, H_2O_2, and ·OH are known to be toxic to most pathogens.

Possible Roles for Zinc in the Oxygen Burst

Those reactions of the "oxygen burst" which might be zinc dependent (Figure 2) will be the focus of this section. The mechanisms of stimulation of NADPH oxidase by PMA and opsonized zymosan are known (Figure 2). PMA stimulates NADPH oxidase by binding to protein kinase C (Nishizuka, 1984) which, in turn, phosphorylates and thus activates NADPH oxidase via arachidonic acid released from phospholipids (20:4) (Broberg and Pick, 1984). In this case, the opsonized zymosan binds to receptors on macrophages for the Fc portion of the antibody (Figure 2). The Fcγ2b binding protein of macrophage membranes has phospholipase A_2 activity resulting in the cleavage of fatty acids such as 20:4 from the C2 position of the glycerol backbone of phospholipids (Suzuki et al., 1982). Free 20:4 and lipoxygenase metabolites of 20:4 such as leukotrienes, 15-HETE or 15-HPETE, stimulate NADPH oxidase (Broberg and Pick, 1984; McPhail et al., 1985). Nonopsopnized zymosan stimulates degradation via both phospholipase A and phospholipase C (Emilsson and Sundler, 1986). Thus, upon binding of opsonized zymosan, phospholipase A_2 or phospholipase C - diacylglycerol lipase liberates 20:4 from membrane phospholipids and 20:4 or a lipoxygenase (but not cyclo-oxygenase) metabolite of 20:4 stimulates NADPH oxidase for the production of O_2-. The PMA stimulated release of 20:4 via phospholipase A (Emilsson and Sundler, 1986) does not activate NADPH oxidase (Figure 2) since, with this stimulant, 20:4 is converted to prostaglandins (Brune et al., 1984) which are not stimulatory for NADPH oxidase (Broberg and

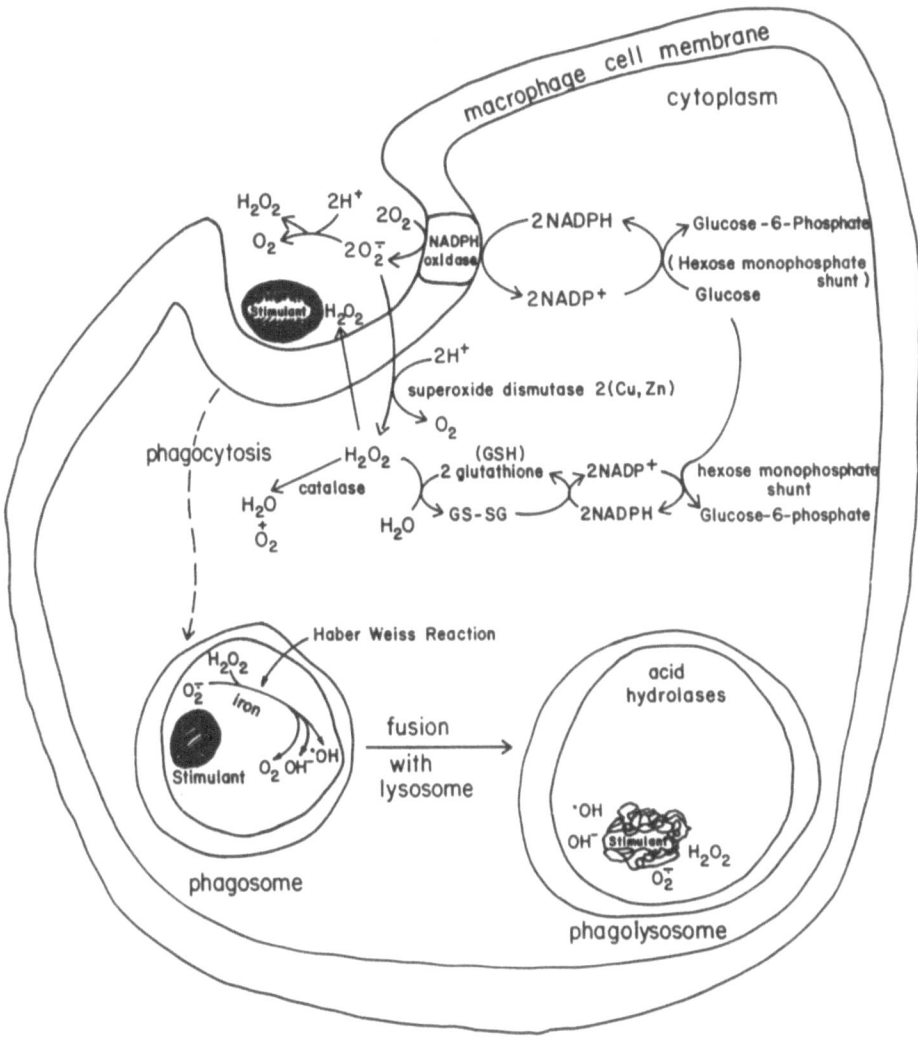

Figure 1. Depiction of the oxygen burst and the production of toxic oxygen metabolites by mononuclear phagocytic cells.

Pick, 1984). Apparently, the particular phospholipases involved in the degradation of phospholipids varies with the type of stimulant encountered. There are several possible roles for zinc in the stimulation of NADPH oxidase by 20:4. Zinc may be important for the release of 20:4 from phospholipids, since phospholipase C is a zinc dependent enzyme (Figure 2) (Dennis, 1983). Phospholipase C has two zinc atoms in the active site, both of which are tightly bound and required for activity (Dennis, 1983). Zinc has also been implicated in the regulation of phospholipase A_2 activity (Horrobin et al., 1978; Manku et al., 1979; Bettger and O'Dell, 1981). Manku et al. (1979) have suggested that in the presence of physiological levels of zinc, dihomogamma-linolenic acid (DHGL), an immediate precursor to 20:4 or prostaglandins, is mobilized from plasma membrane phospholipids of the rat superior mesenteric vascular bed. If DHGL is converted to 20:4, mobilization of DHGL could play a significant role in stimulating the oxygen burst via 20:4. Zinc mobilization of DHGL for prostaglandin synthesis would not be stimulatory for NADPH oxidase. However, inhibition of prostaglandin synthesis may allow DHGL or its metabolite, 20:4, to stimulate NADPH oxidase. This is a possibility since addition of 2mM zinc *in vitro* has been shown to inhibit prostaglandin synthesis by polymorphonuclear leukocytes (Bettger and O'Dell, 1981). In addition, perhaps DHGL itself could stimulate NADPH oxidase since shorter chain fatty acids can, albeit to a lesser extent, also stimulate

NADPH oxidase (Broberg and Pick, 1984). In contrast to the increased activity of phospholipase A_2 in the presence of zinc reported by Manku et al. (1979), Wells (1973) has reported that zinc inhibits the calcium dependent activity of phospholipase A_2 from snake venom by binding to the active site and inducing conformational changes in the enzyme. Inhibition of phospholipase A_2 activity by addition of zinc has also been reported by others (Zor et al., 1969; Stossel et al., 1970). Thus, the effect of zinc on the function of the latter enzyme remains controversial. In summary, zinc may be important in the release of 20:4 from phospholipids via its role in phospholipase C and phospholipase A_2 activity.

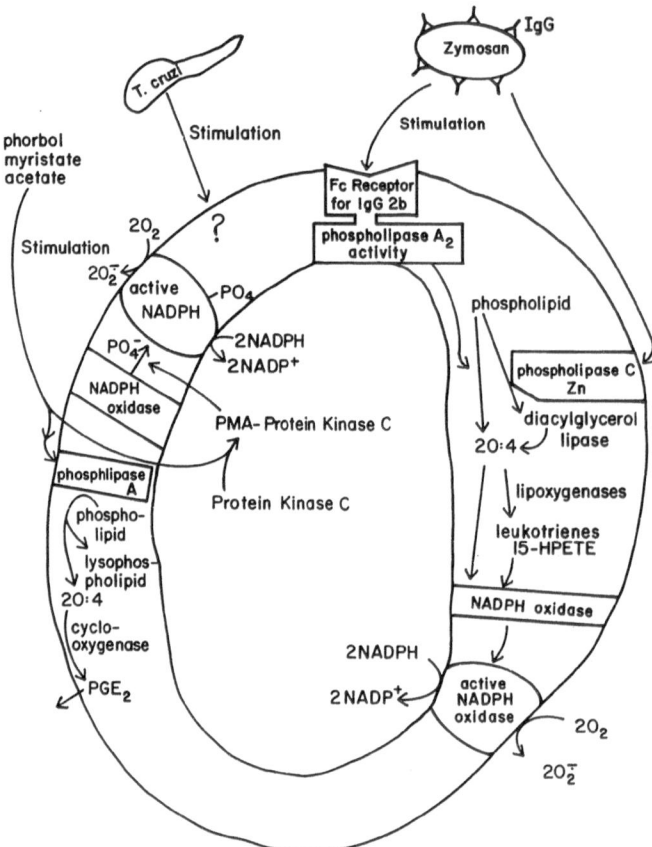

Figure 2. Membrane events associated with initiation of the oxygen burst in mononuclear phagocytes by different stimuli. Examples given here include: the chemical, phorbol myristate acetate; the parasite, *Trypanosoma cruzi*; yeast cell wall fragments, zymosan.

Zinc may also be important in the stabilization of 20:4 against oxidation and thereby enhance the probability of stimulation of NADPH oxidase. It is known that zinc or iron can complex with 20:4 and oxygen (Peterson et al., 1981). Iron catalyzes the oxidation of 20:4 whereas zinc does not (Peterson et al., 1981). These two reactions are as follows: (1) $20:4 + O_2$ $+ FE^{2+} - > [20:4, O_2, FE^{2+}]$ complex $- >$ lipid peroxides $+ O_2- + FE^{3+}$ and (2) $20:4 + O_2 + Zn^{2+}$ $-> [20:4, O_2, Zn^{2+}]$ complex. Thus, zinc may compete with iron for formation of this complex (Peterson et al., 1981) making it unavailable for oxidation by the iron complex and thereby allowing it to act as a stimulant for NADPH oxidase.

Zinc may also play a role in the production of O_2- by NADPH oxidase since nucleotides, such as NADPH, are able to complex with zinc. Zinc binds to NADPH in a 2:1 molar ratio (Slater, 1974). The first zinc atom binds between the monophosphate on the C2 of the adenineribosyl portion and the diphosphate group (Slater, 1974). The second zinc atom binds to the remaining oxygen of the monophosphate and the diphosphate (Slater, 1974). When zinc complexes with NADPH, it interferes with oxidation of NADPH by making it unavailable as a substrate (Slater, 1974). The activity of the NADPH oxidases derived from liver microsomes and mixed function oxidases from smooth endoplasmic reticulum, are inhibited in the presence of zinc (K_i=7.22 μM Zn^{2+}) (Slater, 1974). In the presence of physiological levels of zinc, the importance of the possible formation of the Zn_2-NADPH complex in the oxygen burst is unknown. Conversely, supraphysiological levels of zinc might inhibit NADPH oxidase via complexes formed with NADPH. This might explain the inhibition of oxygen consumption and hexose monophosphate shunt activity in polymorphonuclear neutrophils and MNP when high concentrations of zinc were added to cell preparations in vitro (Chvapil et al., 1977).

Superoxide, produced by NADPH oxidase, either nonenzymatically dismutates to H_2O_2, or superoxide dismutase (SOD) catalyses its dismutation (Figure 1). Both the spontaneous and enzymatic dismutation of O_2^- are dependent on metals (Ricchelli et al., 1983; O'Neill et al., 1983). Cytoplasmic SOD contains both copper and zinc while mitochondrial SOD contains manganese. For cytoplasmic SOD, the copper ions are catalytic cofactors whereas it is thought that the role of zinc in SOD is primarily as a structural cofactor (O'Neill et al., 1983). Thus, zinc appears to not be crucial for activity of the Cu/Zn SOD.

The nonenzymatic dismutation of O_2^- may also be metal dependent. Iron has been shown to react with O_2- and indirect evidence suggests that zinc may also be involved in related reactions. The reaction facilitated by iron is called the Haber-Weiss reaction (Figure 1).

$$Fe^{3+} O_2\text{- --> } Fe^{2+} + O_2$$

$$FE^{2+} + H_2O_2 \text{ --> } Fe^{3+} + OH\text{- } + \cdot OH$$

$$\cdot OH + H_2O_2 \text{ --> } H_2O + HO_2\cdot$$

$$HO_2\cdot + H_2O_2 \text{ --> } O_2 + H_2O + \cdot OH$$

$$\cdot OH + FE^{2+} \text{ --> } FE^{3+} + OH\text{-}$$

The resulting oxygen metabolites are OH- and the toxic ·OH. These reactions occur at the acidic pH's found in the phagosome/phagolysosome which contain the engulfed pathogen or stimulant (Figure 1).

Zinc may also be involved in the nonenzymatic dismutation of O_2^-. There is evidence, albeit inconclusive, that zinc catalyzes reactions with oxygen. In the following reactions, reduction of oxygen in the presence of zinc and bipyridine (bipy) yields highly reactive radicals and zinc peroxides (Sawyer et al., 1984).

$$\overset{-0.5V}{Zn^{II} (bipy)_2{}^{2+} + O_2 + e\text{- -------> } [Zn^{II} (bipy)_3 OO\cdot]+}$$

$$[Zn^{II} (bipy)_2\text{-}OO\cdot]^+ + e\text{- -------> } Zn^{II} (bipy)_2 (O_2)$$

These reactions were determined electrochemically under aprotic conditions using dimethylformamide in order to stabilize the superoxide ion so that it would not dismutate to H_2O_2. For this reason the significance of these reactions for biological systems also remains unknown. In addition, it has also been observed that oxygen reacts with metal surfaces as follows: O_2 (g) -> $O_2^{\cdot}$ (a) -> O- (a) -> O_2^-(a) (Au and Roberts, 1986). Since these reactions were done at extremely low temperatures, the biological relevance is once again unknown. Other metals (iron, copper, manganese, cobalt) are more reactive with oxygen than zinc since they have unfilled d-shell orbitals. However, a zinc metalloporphyrin is reactive with superoxide and forms a superoxo complex (Valentine et al., 1977). These reactions serve to demonstrate that zinc, especially a biochemical complex containing zinc, may in the future prove to have biological reactivity with oxygen species essential to the oxygen burst.

In summary, zinc may participate in a variety of the reactions instrumental to the production of toxic oxygen metabolites by MNP. Zinc may be involved in the stimulation of NADPH oxidase via its role as a cofactor for phospholipase A_2 or phospholipase C - diacylglycerol lipase catalyzed release of 20:4 from phospholipids. Zinc may also stabilize 20:4 against oxidation by iron complexes thereby enhancing the probability that 20:4 will stimulate NADPH oxidase and initiate the oxygen burst. Finally, as yet unidentified biological complexes with zinc may prove to be extremely reactive with oxygen thereby generating products highly toxic to pathogens.

Effects of Zinc Deficiency on H_2O_2

Destruction of *T. cruzi in vivo* (Fraker et al., 1982) and *in vitro* (Table 1) is dependent upon zinc. Moreover, this deficiency is specifically reversed by zinc (Table 1). Killing of *T. cruzi* is thought to involve H_2O_2 production by MNP, albeit much of the evidence correlating H_2O_2 production and killing of *T. cruzi* was actually obtained by indirect means (Nathan et al., 1979; Nathan et al., 1979). That is, elicited MNP were activated by PMA or lipopolysaccharide thereby causing production of H_2O_2 which in turn was shown to be able to kill exogenously added *T. cruzi*. If H_2O_2 is important in killing of *T. cruzi*, perhaps the reduced killing by MNP from zinc deficient mice is due to impaired H_2O_2 production by the macrophages. Direct quantitation of *T. cruzi*-stimulated H_2O_2 production by MNP has been hampered by the lack of assay systems which measure the low amounts of H_2O_2 produced by resident (nonelicited) macrophages stimulated with *T. cruzi*. Additionally, the commonly used phenol red assay for measuring H_2O_2 is toxic to *T. cruzi* (manuscript in preparation). Clearly, this field would be enhanced by improvements in the assay systems. We have modified a H_2O_2 assay which is not toxic to *T. cruzi* and enables one to use actual pathogens as stimulants and quantitate the relatively low amount of H_2O_2 produced by resident MNP (manuscript in preparation).

Table 2. Effect of a Dietary Zinc Deficiency on H_2O_2 Production by Resident Peritoneal Macrophages

Dietary Group	H_2O_2 Production upon Stimulation with				H202/Parasite Treatment
	opsonized zymosan	20:4	PMA	*T. cruzi*	
Severe	100 ± 1^a	74 ± 10	112 ± 2	61 ± 3^b	94 ± 6^c
Moderate	100 ± 1	116 ± 26	144 ± 27	76 ± 3^b	112 ± 6
Restricted	93 ± 1	105 ± 10	96 ± 33	nd^d	nd
Control	100 ± 7	100 ± 21	100 ± 14	100 ± 3	100 ± 6

[a]Percent of control for mean $\pm$ SEM for nmoles H_2O_2/mg macrophage protein produced in 90 minutes.
[b]The difference between this value and the value of the control is statistically different at $p<0.05$ or greater.
[c]Percent of control for nmoles H_2O_2/mg macrophage protein /*T. cruzi*.
[d]nd indicates not determined.

Using this modified assay, we analyzed the ability of resident MNP from zinc deficient mice to produce H_2O_2 upon stimulation with several agents. The amount of zinc present during these studies was minimal (<1 µg Zn/dl). As a comparison, Table 1 shows that 1000 µg $ZnCl_2$/dl (500 µg Zn/dl) was required for restoration of microbicidal activity of deficient MNP. Opsonized zymosan and the second messenger in opsonized zymosan stimulation of H_2O_2 production, 20:4, stimulated the same amount of H_2O_2 production by MNP from zinc deficient and zinc adequate mice (Table 2). Although opsonized zymosan, an antibody coated yeast cell extract, is a more natural stimulant, PMA, a chemical, seems to be the agent most often used to stimulate H_2O_2 production. Upon stimulation with PMA, MNP from zinc deficient mice again produced normal amounts of H_2O_2 in 90 minutes as compared to controls. Therefore, the mechanism for stimulation of H_2O_2 production by opsonized zymosan or PMA was not impaired in MNP from zinc deficient mice.

However, the mechanism for stimulation of NADPH oxidase by *T. cruzi* is unknown and may be different than that for opsonized zymosan or PMA. Further, the mechanism for killing of *T. cruzi* is thought to be dependent upon H_2O_2 and deficiencies in zinc impaired destruction of *T. cruzi* suggesting that *T. cruzi*-stimulated MNP from zinc deficient mice may have a reduced capacity to produce H_2O_2. Upon stimulation with *T. cruzi*, MNP from severely and moderately zinc deficient mice produced significantly less H_2O_2 per macrophage (63% to 73% of control) than zinc adequate MNP (Table 2). However, this was determined to be due to less overall activation of the deficient MNP since fewer parasites had associated with these cells as seen in previous experiments (Table 1). When considered from this point of view, the amount of H_2O_2 produced per parasite was the same for MNP from each dietary group of mice (Table 2). One would expect that destruction of a lower burden of parasites such as that associated with the deficient MNP would require less H_2O_2. With this assumption, the conclusion can be made that the mechanism for production of H_2O_2 is unaltered in the MNP from zinc deficient mice. Therefore, the absence of dietary zinc must alter some other microbicidal process of the MNP since MNP from zinc deficient mice had a reduced ability to destroy those parasites that had associated with it (Table 1).

Still, previous evidence (Nathan et al., 1979; Nathan et al., 1979) and correlations between killing of *T. cruzi* and H_2O_2 production by resident MNP stimulated directly with *T. cruzi* (manuscript in preparation) suggest that H_2O_2 is important in destruction of *T. cruzi*. Perhaps some process in the killing following the production of H_2O_2 requires zinc. However, the actual process of destruction of *T. cruzi* and whether or not some agent in addition to H_2O_2 is important remains to be determined.

CONCLUSIONS

Zinc is an important nutritional trace element for the maintenance of health. The reduced lymphocyte responses by animals is most likely due to reduced numbers of lymphocytes for surveillance since the residual cells seem to retain functional activity. In contrast, MNP, important as a first line of defense against parasitic infections, have an impaired ability to associate with and destroy *T. cruzi*. This impairment is specific for zinc since zinc but not other metals were able to restore the *in vitro* capacity of MNP to destroy *T. cruzi*. H_2O_2 production which is necessary in the destruction of *T. cruzi* by MNP was not impaired in the zinc deficient MNP. Since killing of *T. cruzi* is reduced with the deficiency, perhaps some process in the actual destruction of *T. cruzi* by H_2O_2 or production of some other agent for microbicidal activity requires zinc. Some possible functions for zinc in oxygen metabolism were discussed herein. Biological functions for zinc in the oxygen burst and destruction of pathogens should be the subject of future research.

REFERENCES

Au, C. T., and Roberts, M. W., 1986, Specific role of transient O⁻(s) at Mg(0001) surfaces in activation of ammonia by dioxygen and nitrous oxide, **Nature**, 319:206.

Berg, J., 1986. Potential metal-binding domains in nucleic acid binding proteins, **Science**, 232:485.

Bettger, W. J., and O'Dell, B. L., 1981, A critical physiological role of zinc in the structure and function of biomembranes, **Life Sci.**, 28:1425.

Broberg, Y., and Pick, E., 1984, Unsaturated fatty acids stimulate NADPH-dependent superoxide production by cell free system derived from macrophages, **Cell Immunol.**, 88:213.

Brune, K., Aehringhaus, U., and Peskar, B. A., 1984, Pharmacological control of leukotriene and prostaglandin production from mouse peritoneal macrophages, **Agents Actions**, 14:729.

Chvapil, M., Stankova, L., Berhard, D. S., Weldy, P. L., Carlson, E. C., and Campbell, J. B., 1977, Effect of zinc on peritoneal macrophages *in vitro*, **Infect. Immun.**, 16:367.

Dennis, E. A., 1983, Phospholipases, *in*: **"The Enzymes,"** P. Boyer, ed., Academic Press, New York.

Emilsson, A., and Sundler, R., 1986, Evidence for a catalytic role of phospholipase A in phorbol diester- and zymosan-induced mobilization of arachidonic acid in mouse peritoneal macrophages, **Biochim. Biophy. Acta.**, 876:533.

Fernandes, G., Nair, M., Onoe, K., Tanaka, T., Floyd, R., and Good, R., 1979. Impairment of cell mediated immunity function by dietary zinc deficiency in mice, **Proc. Natl. Acad. Sci.**, 76:457.

Fraker, P. J., Haas, S. M., and Luecke, R. W., 1977, Effect of zinc deficiency on the immune response, **J. Nutr.**, 107:1889.

Fraker, P. J., DePasquale-Jardieu, P., Zwickl, C. M., and Luecke, R. W., 1978. Regeneration of T-cell helper function in zinc deficient adult mice, **Proc. Natl. Acad. Sci.**, 75:5660.

Fraker, P. J., Caruso, R., and Kierszenbaum, F., 1982, Alteration of the immune and nutritional status of mice by synergy between zinc deficiency and infection with *Trypanosoma cruzi*, **J. Nutr.**, 112:1224.

Fraker, P. J., Zwickl, C. M., and Luecke, R. W., 1982, Delayed type hypersensitivity in zinc deficient adult mice, **J. Nutr.**, 111:409.

Fraker, P. J., Gershwin, M., Good, R., and Prasad, A., 1986, Interrelationships between zinc and immune function, **Fed. Proc.**, 45:1474.

Fraker, P. J., Jardieu, P., and Cook, J., 1987, Zinc deficiency and immune function, **Arch. Int. Med.**, 147:1699.

Fraker, P. J., Jardieu, P., and Cook, J. M., 1988, Immunodeficiency caused by an inadequate intake of zinc, *in*: **"The Nature, Cellular and Biochemical Basis and Management of Immunodeficiencies,"** Lindenlaub, ed., Symposia Medica Hoechst, Frankfort.

Gershwin, M.E., Beach, R. S., and Hurley, L. S., 1979. Trace elements, *in*: **"Nutrition and Immunity,"** Academic Press, New York.

Good, R. A., 1981, Nutrition and immunity, **J. Clin. Immunol.**, 1:3.

Horrobin, D. F., Manku, M. S., Cunnane, S., Karmazyn, M., Morgan, P. O., Ally, A. I., and Karmall, R. A., 1978. Regulation of cytoplasmic calcium: Interactions between prostaglandins, prostacyclin, thromboxane A_2, zinc, copper and taurine, **Can. J. Neur. Sci.**, 5:93.

Manku, M. S., Horrobin, D. F., Karmazyn, M., and Cunnane, S. C, 1979, Prolactin and zinc effects on rat vascular reactivity: Possible relationship to dihomo-τ-linolenic acid and to prostaglandin synthesis, **Endocrinology**, 104:774.

McPhail, L. C., Shirley, P. S., Clayton, C. C., and Snyderman, R., 1985, Activation of the respiratory burst enzyme from human neutrophils in a cell-free system, **J. Clin. Invest.**, 75:1735.

Nathan, C., Nogueira, N., Juangbhanich, C., Ellis, J., and Cohn, Z., 1979, Activation of macrophage *in vivo* and *in vitro*: Correlation between hydrogen peroxide release and killing of *Trypanosoma cruzi*, **J. Exp. Med.**, 149:1056.

Nathan, C. F., Silverstein, S. C., Brukner, L. H., and Cohn, Z., 1979, Extracellular cytolysis by activated macrophages and granulocytes. II. Hydrogen peroxide as a mediator of cytotoxicity, **J. Exp. Med.**, 149:100.

Nishizuka, Y., 1984, The role of protein kinase C in cell surface signal transduction and tumor promotion, **Nature**, 308:693.

O'Neill, P., Fielden, E. M., Cocco, D., Calabrese, L., and Rotillo, G., 1983, Mechanistic study of superoxide dismutation by "zinc-free" bovie superoxide dismutase, *in*: **"Oxy Radicals and Their Scavenger Systems,"** W. Gors, M. Saran, and D. Tait, eds., Elsevier Science Publishing Co., New York.

Peterson, D. A., Gerrard, J. M., Peller, J., Ras, G.H.R., and White, J. G., 1981, Interactions of zinc and arachidonic acid, **Prostaglandins Med.**, 6:91.

Prasad, A., 1979, Clinical, biochemical, and pharmacological role of zinc, **Ann. Rev. Pharmacol.**, 20:393.

Prasad, A., 1984, Discovery and importance of zinc in human nutrition, **Fed. Proc.**, 43:2829.

Ricchelli, F., Rossi, E., Salvato, B., Jori, G., Bannister, J. V., and Bannister, W. H., 1983, Fluorescence studies on copper/zinc superoxide dismutase from bovine erythrocytes, *in*: **"Oxy Radicals and Their Scavenger Systems,"** G. Cohen and R. A. Greenwald, eds, Elsevier Science Publishing Co., New York.

Sandstead, H., 1973, Zinc nutrition in the United States, **Am. J. Clin. Nutr.**, 26:1251.

Sawyer, D. T., Roberts, J. L. Fr., Tsuchiya, T., and Srivatsa, G. S., 1984, Generation of activated oxygen species by electron-transfer reduction of dioxygen in the presence of protons, chlorinated hydrocarbons, methyl viologen and transition metal ions, *in*:

"Oxygen Radicals in Chemistry and Biology," W. Gors, M. Saran, and D. Tait, eds, Walter de Gruyter & Co., New York.

Slater, T. F., 1974, Mechanisms of protection against the damage produced in biological systems by oxygen-derived radicals, *in*: "Molecular Mechanisms of Oxygen Activation," Q. Hayaishi, ed., Academic Press, New York.

Stossel, T. P., Murad, F., Mason, R. J., and Vaughan, M., 1970, Regulation of glycogen metabolism in polymorphonuclear leukocytes, **J. Biol. Chem.**, 245:6228.

Suzuki, T., S aito-Taki, T., Sadasivan, R., and Nitta, T., 1982, Biochemical signal transmitted by $Fc\tau$ receptors: Phospholipase A_2 activity of $Fc\tau2b$ receptor of murine macrophage cell line P388D, **Proc. Natl. Acad. Sci. USA.**, 79:591.

Valentine, J. S., Tatsuno, Y., and Nappa, M., 1977, Super oxotetraphenylporphinatozinc (1-), **J. Am. Chem. Soc.**, 99:3522.

Vallee, B., and Galdes, A., 1984, The metallobiochemistry of zinc enzymes, **Adv. Enzymol.**, 56:283.

Wells, M. A., 1973, Spectrol perturbations of *Crotalus adamanteus* phospholipase A_2 induced by divalent cation binding, **Biochem.**, 12:1080.

Wirth, J. J., Fraker, P. J., and Kierzenbaum, F., 1984, Changes in the levels of marker expression by mononuclear phagocytes in zinc deficient mice, **J. Nutr.**, 114:1826.

Zor, U., Kaneko, T., Lowe, I. P., Bloom, G., and Field, J., 1969, Effect of thyroid-stimulating hormone and prostaglandins on thyroid adenyl cyclase activation and cyclic adenosine 3',5'monophosphate, **J. Biol. Chem.**, 244:5189.

EFFECTS OF COPPER DEFICIENCY ON THE IMMUNE SYSTEM

Joseph R. Prohaska and Omelan A. Lukasewycz

Departments of Biochemistry and Medical Microbiology and Immunology
University of Minnesota, Duluth
Duluth, MN 55812

INTRODUCTION

Copper is an essential metal for proper functioning of all living systems. Biochemical mechanisms have evolved that result in homeostatic balance of copper. This ensures that adequate but not toxic levels are absorbed, transported, utilized, and excreted. Throughout the biological kingdom copper expresses its function through specific ligands as free copper ion is rapidly complexed. These ligands are usually specific cuproenzymes. Knowledge of these cuproenzymes forms the basis of our current understanding of the biochemical function of copper (Prohaska, 1988).

In animals there are approximately ten proteins that are generally accepted as true cuproenzymes (Prohaska, 1988). They range in size from small single polypeptides to complex tetramers with up to eight Cu atoms per mole. Some cuproenzymes have additional cofactor requirements such as zinc, heme iron, or pyrroloquinoline quinone. It is this diverse nature of the many cuproenzymes that is responsible for the variety of specialized copper functions.

In addition to the ten established cuproenzymes there is a group of 12 proteins that when isolated contain one or more Cu atoms. The biological function of these cuproproteins is not known with certainty (Prohaska, 1988). Some of these cuproproteins may have enzyme functions discovered and become established as true cuproenzymes. However further purification often results in elimination of putative cuproenzymes as was the fate of seven such proteins (Prohaska, 1988).

The direct effects of copper are well established. When environmental copper levels become limiting or excessive the activity of a number of other enzymes is known to change, sometimes higher more often lower. A recent review cited sixteen such enzymes (Prohaska, 1988). The indirect effects of copper also influence biological functioning.

The ability to withstand an invasion and develop immunity against a bacterial or viral pathogen requires a complex interplay between multiple organs, cell types and molecules. It is, thus, not surprising that copper is one such factor in host defense mechanisms. Evidence that copper plays an important role in host defenses has come from a variety of sources including environmental copper deficiency in humans and domestic animals, genetic copper deficiency in humans (Menkes' disease), and dietary copper deficiency in experimental rodents. From this evidence several scenarios can be presented to explain the altered immunity and increased infections that occur when copper is limiting, each involving one or more cuproenzyme (Figure 1). Altered antioxidant status is one such scenario since two cuproenzymes, ceruloplasmin and Cu,Zn-superoxide dismutase (Cu,Zn-SOD), are known to exhibit antioxidant function. Another possibility involves altered energy metabolism due to limiting cytochrome c oxidase. A third possible scenario involves altered neuroendocrine function. The synthesis of the neurotransmitter norepinephrine depends on the cuproenzyme dopamine-β-monooxygenase and the formation of several bioactive peptides known to influence the immune response such as vasoactive intestinal peptide (VIP) depend on another cuproenzyme, peptidyl-glycine α-amidating monooxygenase. Copper may also be involved in intra and intermolecular disulfide formation in proteins such as immunoglobulins. These alternative hypotheses will be discussed later in the paper.

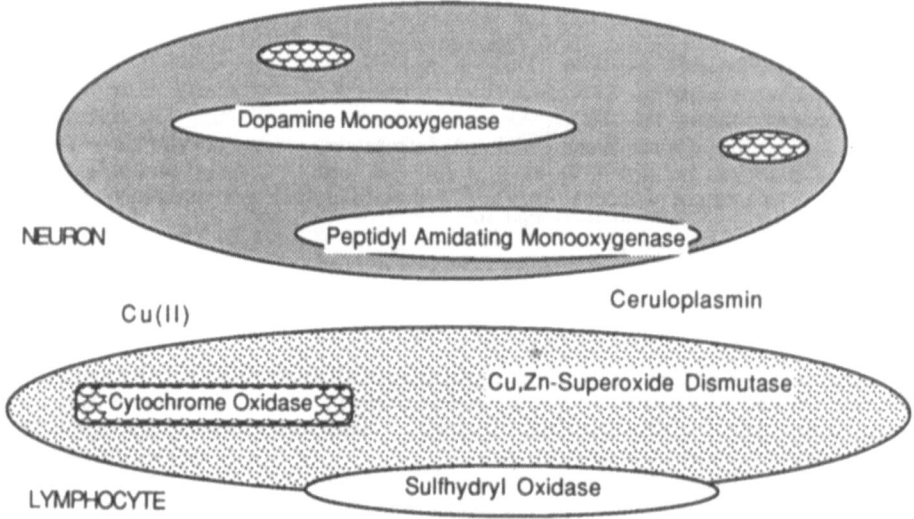

Figure 1. Cuproenzymes Involved in the Neuroimmune Response.

Human Copper Deficiency

Copper is an essential nutrient for biological systems including the development and maintenance of the immune system. Although currently difficult to diagnose, marginal copper deficiency may exist in the human population. It is probable that adequate dietary copper is important in expressing full immune function in humans.

Until recently there has been little concern regarding copper deficiency in humans. However, Sandstead (1982) has suggested that modest copper deficiency is evident even in "healthy" adult populations. Furthermore, the incidence of this modest deficiency of copper can be expected to increase because of negative dietary interactions with protein, fiber, zinc and ascorbate. For example, reports continue of adults that develop anemia and neutropenia from "megadosing" with zinc (Simon et al., 1988). Acquired severe copper deficiency in adults, although rare, develops occasionally as in the case of a 76 year old female that presented with anemia, neutropenia, and hypogammaglobulinemia (Oppenheimer et al., 1987)

However, the most susceptible population to copper deficiency are infants. Gibson (1985) has reported that the intake of copper by breast-fed or formula-fed infants is below the US RDA. Furthermore, the intake of copper following weaning may be hampered by poor absorption from cereals (Bell et al., 1987) and by negative interactions with supplements such as iron (Haschke et al., 1986). During infancy, copper deficiency is often associated with infection as documented in seven separate reports (Table 1). A most provocative study completed in Chile (Castillo-Duran et al., 1983) suggested that copper supplementation was needed in treating marasmic infants during recovery to prevent severe respiratory infections. Many of the unsupplemented children showed only marginal signs of copper deficiency!

1. Al-Rashid, R. A., and Spangler, J., 1971, Neonatal copper deficiency, **New Engl. J. Med.**, 285:841-843.
2. Karpel, J. T., and Peden, V. H., 1972, Copper deficiency in long-term parenteral nutrition, **J. Pediatrics**, 80:32-36.
3. Sann, L., David, L., Galy, G., and Romand-Monier, R., 1978, Copper deficiency and hypocalcemic rickets in a small-for-date infant, **Acta Paediatr. Scand.**, 67:303-307.
4. Yuen, P., Lin, H. J., and Hutchison, J. H., 1979, Copper deficiency in a low birthweight infant, **Arch. Dis. Child.**, 54:553-554.
5. Bennani-Smires, C., Medina, J., and Young, L. W., 1980, Radiological case of the month, **Am. J. Dis. Child.**, 134:1155-1156.
6. Allen, T. M., Manoli, A., and Lamont, R. L., 1982, Skeletal changes associated with copper deficiency, **Clin. Orthoped. Rel. Res.**, 148:206-210.
7. Levy, J., Berdon, W. E., and Abramson, S. J., 1984, Epiphyseal separation simulating pyarthrosis, secondary to copper deficiency, in an infant receiving total parenteral nutrition, **Br. J. Radiol.**, 57:636-638.

Copper deficiency can also occur in Menkes' syndrome, a lethal X-linked abnormality. Cause of death in most reported cases is bronchopneumonia (Danks et al., 1972). Between 1966 and 1985 twenty-nine separate reports have been published which describe patients with Menkes' disease having infection with cause of death as bronchopneumonia (Table 2).

It is true that inactivity in these sick children predisposes them to respiratory infections but seven other reports about children with Menkes' disease also discuss frequent infections (Table 3). Thus, copper nutriture is critical during the time the immune system is developing. Copper is definitely required for maturation and formation of erythrocytes and neutrophils as anemia and neutropenia are common clinical observations in copper-deficient infants. It is likely that copper is also required for lymphocyte development.

Animal Copper Deficiency

Observations of domestic animals and research with experimental rodents has confirmed and extended the evidence that copper is essential for host defenses. For example, *Mycoplasma* and *Haemophilus* infections were noted as primary causes of death in copper-deficient (-Cu) domestic piglets (Pletcher and Banting, 1983). Increased mortality to challenge by *Salmonella typhimurium* was noted in -Cu rats twenty years ago by Newberne and colleagues (1968). Studies with -Cu mice showed increased mortality to challenge by *Pasteurella hemolytica* (Jones and Suttle, 1983). Also -Cu mice showed high mortality to an immunizing dose of syngeneic malignant lymphocytes (Lukasewycz and Prohaska, 1982). It seems quite clear that the copper-deficient state in humans and in animals is associated with increased risk of infection.

What are the immunological and biochemical mechanisms? Studies with C57BL mice are presented that illustrate biochemical and immunological changes following dietary copper deficiency during perinatal development and following genetic copper deficiency in a murine mutant analog of Menkes' disease, the brindled mouse.

METHODS AND MATERIALS

Animal Care and Diets

Mice were maintained in our colony. They were derived from pairs kindly provided by Dr. David Grahn, Argonne National Laboratory. They were from a C57BL strain and the original females were heterozygous for the brindled gene at the mottled locus of the X-chromosome designated as follows: +/Mo^{br}. Two month old male and female mice were mated to establish a breeding colony. Two dietary treatments were employed in these experiments, copper-adequate (+Cu) and copper-deficient (-Cu). The -Cu treatment consisted of feeding a semipurified diet that omitted cupric carbonate from the salt mix (modified AIN-76A, Teklad Laboratories, Madison, WI). By analysis the diet contained 0.43 ± 0.042 [6] mg/kg copper

Table 2. Case Reports of Menkes' Disease in Which Cause of Death was Bronchopneumonia.

1. Aguilar, M. J., Chadwick, D. L., Okuyama, K., and Kamoshita, S., 1966, Kinky hair disease. I. Clinical and pathological features, **J. Neuropathol. Exp. Neurol.**, 25:507-522.
2. French, J. H., Sherard, E. S., Lubell, H., Brotz, M., and Moore, C. L., 1972, Trichopoliodystrophy, **Arch. Neurol.**, 26:229-244.
3. Danks, D. M., Campbell, P. E., Stevens, B. J., Mayne, V., and Cartwright, E., 1972, Menkes' kinky hair syndrome. An inherited defect in copper absorption with widespread effects, **Pediatr.**, 50:188-201.
4. Dorn, G., Neuhäuser, G., Heye, D., and Kielhorn, A., 1973, Das kinky-hair syndrom von Menkes, **Klin. Padiat.**, 185:480-489.
5. Mollekaer, A. M., 1974, Kinky hair syndrome, **Acta Paediat. Scand.**, 63:289-296.
6. Grover, W. D., and Scrutton, M. C., 1975, Copper infusion therapy in trichopoliodystrophy, **J. Pediatr.**, 86:216-220.
7. Wheeler E. M., and Roberts, P. F., 1976, Menkes's steely hair syndrome, **Arch. Dis. Child.**, 51:269-274.
8. Oakes, B. W., Danes, D. M., and Campbell, P. E., 1976, Human copper deficiency: ultrastructural studies of the aorta and skin in a child with Menkes' syndrome, **Exp. Mol. Pathol.**, 25:82-98.
9. Hirano, A., Llena, J. F., French, J. H., and Ghatak, N. R., 1977, Fine structure of the cerebellar cortex in Menkes kinky-hair disease, **Arch. Neurol.**, 34:52-56.
10. Hockey, A., and Masters, C. L., 1977, Menkes' kinky (steely) hair disease, **Aust. J. Derm.**, 18:77-80.
11. Garnica, A. D., Frias, J. L., and Rennert, O. M., 1977, Menkes kinky hair syndrome: is it a treatable disorder? **Clin. Genet.**, 11:154-161.
12. Martin, J. J., Flament-Durand, J., Farriaux, J. P., Buyssens, N., Ketelbant-Balasse, P., and Jansen, C., 1978, Menkes kinky-hair disease: a report on its pathology, **Acta Neuropathol.**, 42:25-32.
13. Williams, R. S., Marshall, R. C., Lott, I. T., and Caviness, V. S., 1978, The cellular pathology of Menkes steely hair syndrome, **Neurol.**, 28:575-583.
14. Grover, W. D., Johnson, W. C., and Henkin, R. I., 1979, Clinical and biochemical aspects of trichopoliodystrophy, **Ann. Neurol.**, 5:65-71.
15. Iwata, M., Hirano, A., and French, J. H., 1979, Thalamic degeneration in X-chromosome-linked copper malabsorption, **Ann. Neurol.**, 5:359-366.
16. Hara, K., Oohira, A., Nogami, H., Watanabe, K., and Miyazaka, S., 1979, Kinky hair disease. Biochemical, histochemical, and ultrastructural studies, **Pediatr. Res.**, 13:1222-1226.
17. Ratanarapee, S., and Tuchinda, C., 1981, Trichopoliodystrophy: report of a case with ultrastructural study, **J. Med. Assoc. Thailand**, 65:158-166.
18. Nooijen, J. L., DeGroot, C. J., van den Hamer, C. J. A., Monnens, L. A. H., Willemse, J., and Niermeijer, M. F., 1981, Trace element studies in three patients and a fetus with Menkes' disease. Effect of copper therapy, **Pediatr. Res.**, 15:284-289.
19. Taylor, C. J., and Green, S. H., 1981, Menkes' syndrome (trichopoliodystrophy): use of scanning electron-microscope in diagnosis and carrier identification, **Dev. Med. Child Neurol.**, 23:361-368.
20. Loyola, M. A., and Dodson, N. E., 1981, Consequences of trichopoliodystrophy, **J. Pediatr.**, 98:588-591.
21. Willemse, J., van den Hamer, C. J. A., Prins, H. W., and Jonker, P. L., 1982, Menkes' kinky hair disease. I. Comparison of classical and unusual clinical and biochemical features in two patients, **Brain Dev.**, 4:105-114.
22. Troost, D., van Rossum, A., Straks, W., and Willemse, J., 1982, Menkes' kinky hair disease. II. A clinicopathological report of three cases, **Brain Dev.**, 4:115-126.
23. Yoshimura, N., and Kudo, H., 1983, Mitochondrial abnormalities in Menkes' kinky hair disease (MKHD), **Acta Neuropathol.**, 59:295-303.
24. Uno, H., Arya, S., Laxova, R., and Gilbert, E. F., 1983, Menkes' syndrome with vascular and adrenergic nerve abnormalities, **Arch. Pathol. Lab. Med.**, 107:286-289.
25. Maehara, M., Ogasawara, N., Mizutani, N., Watanabe, K., and Suzuki, S., 1983, Cytochrome c oxidase deficiency in Menkes kinky hair disease, **Brain Dev.**, 5:533-540.
26. Tan, N., and Urich, H., 1983, Menkes' disease and swayback. A comparative study of two copper deficiency syndromes, **J. Neurol. Sci.**, 62:95-113.
27. Garnica, A. D., 1984, The failure of parenteral copper therapy in Menkes kinky hair syndrome, **Eur. J. Pediatr.**, 142:98-102.
28. Maddox, J. L. Jr., 1984, Menkes's syndrome, **Pediatr. Dermatol.**, 1:307-311.
29. Moore, C. M., and Howell, R. R., 1985, Ectodermal manifestations in Menkes disease, **Clin. Genet.**, 28:532-540.

(Cu) [mean ± SD (N)] and 50.9 ± 1.19 mg/kg iron (Fe). Offspring and dams on the -Cu treatment drank deionized water which contained 0.2 ng/L Cu by analysis. The diet contained the following major components (g/kg): sucrose (500), casein (200), cornstarch (150), corn oil

(50), cellulose (50), mineral mix (35) and vitamin mix (10). During periods when the -Cu treatment was employed control +Cu groups were studied simultaneously by adding Cu to the drinking water (20 mg/L as 80 mg $CuSO_4 \cdot 5H_2O/L$). At other times mice were maintained on another +Cu treatment which consisted of a nonpurified diet (Purina Mouse Chow, Ralston Purina, St. Louis, MO) that contained 15.2 ± 1.42 [6] mg/kg Cu and 118 ± 36.2 mg/kg Fe and tap water which contained 20 ng/L of Cu by analysis. Two days following parturition litter size was adjusted to 8 pups.

Table 3. Case Reports of Menkes' Disease Which Document Frequent Infections.

1. Singh, S., and Bresnan, M. J., 1973, Menkes kinky-hair syndrome (trichopoliodystrophy), **Am. J. Dis. Child.**, 125:572-578.
2. Bucknall, W. E., Haslam, R. H. A., and Holtzman, N. A., 1973, Kinky hair syndrome: response to copper therapy, **Pediatrics**, 52:653-657.
3. Lucky, A. W., and Hsia, Y. E., 1979, Distribution of ingested and injected radiocopper in two patients with Menkes' kinky hair disease, **Pediatr. Res.**, 13:1280-1284.
4. Sakano, T., Okuda, N., Yoshimitsu, K., Hatano, S., Nishi, Y., Tanaka, T., and Usui, T., 1982, A case of Menkes syndrome with cataracts, **Eur. J. Pediatr.**, 138:357-358.
5. Gunn, T. R., Macfarlane, S., and Phillips, L. I., 1984, Difficulties in the neonatal diagnosis of Menkes' kinky hair syndrome--trichopoliodystrophy, **Clin. Pediatr.**, 23:514-516.
6. Blackett, P. R., Lee, D. M., Donaldson, D. L., Fesmire, J. D., Chan, W. Y., Holcombe, J. H., and Rennert, O. M., 1984, Studies of lipids, lipoproteins, and apolipoproteins in Menkes' disease, **Pediatr. Res.**, 18:864-870.
7. Farrelly, C., Stringer D. A., Daneman, A., Fitz, C. R., and Sass-Kortsak, A., 1984, CT manifestations of Menkes' kinky hair syndrome (trichopoliodystrophy), **J. de l'Association Canadienne des Radiologists**, 35:406-408.

Two experimental designs were employed, one studying dietary Cu deficiency and the second genetic Cu deficiency. Induction of dietary Cu deficiency was based on a protocol described previously (Prohaska et al., 1983). During gestation dams were fed the nonpurified diet and starting the day of parturition dams were switched to either the +Cu or -Cu system. Mice were weaned at 18-19 days of age and males were transferred to stainless steel cages at 21 days of age and kept on the respective treatment of their dams. Postmortem data was collected from seven-week old mice. For purposes of analysis, data from several representative experiments were pooled.

Genetic copper deficiency was studied in two experiments. In the first, adult ten month old female brindled mice, $+/Mo^{br}$, were compared to age-matched control mice, $+/+$. In the other 12-day old male offspring of brindled dams were compared to male offspring of control dams. The brindled dams were fed the nonpurified diet and the control dams were placed on either the +Cu or -Cu treatment throughout gestation and lactation as in previous studies (Prohaska, 1983). Two genotypes were studied, control $+/y$, and brindled Mo^{br}/y.

Blood samples were drawn into heparinized microhematocrit tubes under light ether anesthesia from the retro-orbital plexus or, in the case of the suckling mice, following decapitation. Older mice were killed by cervical dislocation and liver, spleen, and thymus were removed, weighed and processed for biochemical or immunological analysis. In some cases other organs were removed for measurement of Cu and Fe.

Biochemical Analyses

Livers, other organs, and 1 g portions of diets were wet-digested with 4 ml of HNO_3 (AR select grade, Mallinckrodt, St. Louis, MO), and the residue was brought to 4.0 ml with 0.1 N HNO_3. Samples were then analyzed for total copper and iron by flame atomic absorption spectroscopy (Model 2380, Perkin-Elmer, Norwalk, CT). Analyses were checked with a certified standard, U.S. National Bureau of Standards (NBS) 1577 bovine liver.

A 5-μl aliquot of blood was used to determine hemoglobin spectrophotometrically as metcyanohemoglobin at 540 nm by using Drabkin's reagent (Prohaska, 1983). Mouse plasma obtained from the microhematocrit tubes was used to determine ceruloplasmin (EC 1.16.3.1) activity using *o*-dianisidine as substrate at 37° C in 0.1 M sodium acetate (pH 5.5) containing 10 µM diethylenetriaminepentaacetic acid (DETAPAC) (Prohaska, 1983).

A sample of the splenocytes and macrophages not used for immunological assays (~ 5 x 10^7 cells) were processed for enzyme assay. Cells were suspended at room temperature for 5 min in 10 ml of 0.02 M Tris (pH 7.2) containing 0.155 M NH$_4$Cl to lyse erythrocytes. Cells were harvested by centrifugation at 400 x *g* for 10 min. The pellet was suspended (1.5 x 10^8 cells/ml) in 0.05 M potassium phosphate (pH 7.0) containing 0.1% Triton X-100 and mixed thoroughly with a syringe. Following centrifugation, 6,000 x *g* for 10 min, the supernate was used to assay Cu,Zn-SOD (EC 1.15.1.1) activity by following inhibition of pyrogallol autoxidation at 320 nm (Prohaska, 1983). Cells were treated with ethanol-chloroform to inactivate manganese SOD. In other cases whole spleens, thymuses, and aliquots of liver were homogenized in 9 volumes of 0.05 M potassium phosphate (pH 7.0) for further enzyme analysis. Cytochrome *c* oxidase activity was determined spectrophotometrically by following oxidation of ferrocytochrome *c* at 550 nm as described previously (Prohaska, 1983). For this assay tissues are diluted in 0.9% Tween-80 just prior to analysis. Portions of the cell supernate and homogenate were used to determine total protein by a modified Lowry procedure with bovine albumin as a standard (Markwell et al., 1978).

Other liver enzymes were determined to assess antioxidant status. Activity of glutathione peroxidase (GSH-Px) was determined by a coupled enzyme assay using 0.25 mM *t*-butylhydroperoxide as substrate (Prohaska et al., 1988b). Mouse liver glutathione transferases were assessed by measuring the ability of crude liver extracts to conjugate GSH and 1-chloro-2,4-dinitrobenzene (Habig et al., 1974). Catalase activity was determined by following loss of hydrogen peroxide at 240 nm and 25° in 0.05 M potassium phosphate (pH 7.5). Velocity measurements made during the initial minute of reaction were converted to units (µmol/min) using a value of 43.6 for molar absorptivity.

Immunological Analyses

Single cell suspensions of splenocytes were prepared as described previously with slight modification (Lukasewycz and Prohaska, 1983). After aseptic removal and mincing the preparation was drawn into a 3 ml syringe with 19 gauge needle for mixing and then transferred to a 15 ml centrifuge tube. The debris was allowed to settle for 4 min and the resulting supernate was used for future evaluation of splenocyte function.

The protocol for lymphocyte culture and reactivity to mitogens has been described in detail (Lukasewycz and Prohaska, 1983). Briefly, 5 x 10^5 splenocytes were cultured in triplicate in the absence or presence of mitogens, pulsed with ^{3}H-thymidine, and analyzed for DNA content by counting in Ecolite (InterChem Enterprises, Inc., San Diego, CA) by liquid scintillation (Model LS 8000, Beckman Corp., Fullerton, CA). The mitogens were purchased commercially (Sigma Chemical Co., St. Louis, MO) and used in the following concentrations at optimal doses: Concanavalin A (Con A) Type IV (0.5 µg/ml); lipopolysaccharide (LPS) *E. coli* 0127:B8 (20 µg/ml); pokeweed (PWM) (10 µg/ml); and phytohemagglutinin-L (PHA) (10 µg/ml). Mean cpm were determined for the triplicate wells and stimulation indices (SI) were calculated by dividing cpm in the presence of mitogen by cpm in the absence of mitogen. Thymocyte cultures were run in an analogous manner using only Con A and PHA as mitogens and 10^6 cells/well.

The number of antibody (IgM) producing cells to sheep red blood cells (SRBC) was determined by *in vitro* culture techniques using a modified system originally developed by Mishell and Dutton (1967). Briefly, 1 x 10^7 splenocytes were cultured in duplicate in the presence of an equal number of SRBC (Kroy Medical, Stillwater, MN) that had been washed 3 times with Hank's balanced salt solution (HBSS) in 10 x 35 mm plastic culture dishes (Corning, Corning, NY). After 4 days of culture, cells were harvested, pooled, and washed with HBSS. Aliquots of the final 2 ml suspension were used to determine viability by Trypan blue exclusion and to enumerate the number of plaque forming cells (PFC) by incubation in triplicate with SRBC in monolayer plates coated with poly-L-lysine. Plates were incubated with guinea pig complement (Anderson Laboratories, Inc., Fort Worth, TX) at 37° C (without CO$_2$)

for 1 h. Hemolytic plaques were counted and data from triplicate cultures averaged, corrected for viability and expressed per original culture (i.e., 2 x 10^7 splenocytes).

Statistical Analyses

Data were analyzed using a Macintosh personal computer and statistical software (Statview 512+, Brain Power, Inc., Calabasas, CA). Tabular results are reported as mean ± SD or SEM. Overall treatment effects were determined by one-way ANOVA or, for two groups, by Student's t-test (α = 0.05 and 0.01).

RESULTS AND DISCUSSION

When dietary copper deficiency was investigated in suckling mice whose dams were on copper-deficient (-Cu) treatment throughout gestation and lactation the offspring did not survive the lactational period (Prohaska and Lukasewycz, 1981; Prohaska, 1983). The size of primary and secondary lymphoid tissue was much lower in deficient pups compared to controls, however, immunological evaluation of those residual organs (spleen and thymus) was not possible in such young mice. Thus, a system was developed to induce dietary Cu deficiency starting at birth (Prohaska and Lukasewycz, 1981). These studies were conducted in an inbred strain of mice, C58, that were used for a specific leukemic model (Lukasewycz and Prohaska, 1982). To confirm and extend those studies dietary copper deficiency was produced in another mouse strain, C57BL, which also was used to study genetic copper deficiency.

Biochemical Changes Following Dietary Copper Deficiency

Male C57BL mice were analyzed at 7-weeks of age following dietary Cu deprivation. Pooled data over a period of two years from the -Cu mice demonstrated signs consistent with moderate Cu deficiency (Table 4). In C57BL mice growth was not altered by the -Cu dietary treatment but organ weights were. In particular, compared to +Cu controls, -Cu mice had higher heart and small intestine weights but lower brain and kidney weight. Since body weight was unaltered, relative organ weight (organ weight/body weight) would have demonstrated the same trends (data not shown). Following the 7-week -Cu treatment period the -Cu mice had ceruloplasmin (CP) activities near the lower limit of detection, less than 2% of +Cu values (Table 4). The -Cu mice were anemic as both hemoglobin and hematocrit demonstrated. Thus, many of the features of the -Cu C57BL mice agree with earlier studies of C58 mice (Prohaska et al., 1983).

Table 4. Effect of Dietary Copper Deficiency from Birth on 7-week-old Male C57BL Mice.

Parameter	Cu-Adequate	Cu-Deficient
Body Wt (g)	22.2 ± 2.26 (30)	21.0 ± 2.29 (30)
Heart Wt (mg)	110 ± 7.45 (8)	154 ± 26.8 (14)[*]
Brain Wt (mg)	433 ± 7.8 (6)	414 ± 14.5 (8)[*]
Kidney Wt (mg)	410 ± 26.6 (6)	340 ± 44.6 (8)[*]
Small Intestine (mg)	211 ± 34.2 (8)	282 ± 53.3 (12)[*]
Hemoglobin (g/100 ml)	15.9 ± 0.75 (12)	12.8 ± 2.9 (12)[*]
Hematocrit (%)	51 ± 0.84 (15)	31.5 ± 9.2 (17)[*]
Ceruloplasmin (units/L)	50.8 ± 5.95 (25)	0.91 ± 2.0 (30)[*]

Values are means ± SD (N). Treatment differences were tested by Student's t-test, * P < 0.01.

Organ levels of Cu and Fe were determined by flame AAS to assess the degree of Cu deficiency in the -Cu mice compared to +Cu controls (Table 5). All five organs examined showed that dietary Cu deficiency resulted in lower Cu levels in organs from -Cu mice compared to +Cu mice; however, the degree of Cu reduction (% +Cu) was variable in the following order, most deficient to least deficient: brain (25% +Cu), heart (28%), small intestine (33%), liver (40%), and kidney (66%). Three organs demonstrated alterations in Fe concentration. Liver of -Cu mice had 227% higher Fe levels than +Cu mice (Table 5). Heart Fe levels of -Cu mice were 76% of that of +Cu mice and are probably modestly low even though heart Fe was not corrected for blood Fe present as a contaminant. Brain Fe levels of -Cu mice were 92% of +Cu values and this small difference is likely due to differences in blood Fe present as a contaminant in the unperfused organs.

Table 5. Copper and Iron Levels in Organs from 7-week-old Male C57BL Mice Following Dietary Copper Deficiency.

	Copper (µg/g)		Iron (µg/g)	
Organ	+Cu	-Cu	+Cu	-Cu
Liver	4.76 ± 0.42 (30)	1.90 ± 0.56 (26)[*]	130 ± 28.8 (30)	295 ± 86.6 (26)[*]
Heart	5.80 ± 0.28 (8)	1.60 ± 0.35 (13)[*]	80.9 ± 5.31 (8)	61.3 ± 8.24 (13)[*]
Brain	3.59 ± 0.18 (6)	0.90 ± 0.23 (8)[*]	16.1 ± 0.46 (6)	14.8 ± 0.92 (8)[*]
Kidney	4.69 ± 0.32 (6)	3.10 ± 0.34 (8)[*]	60.4 ± 5.24 (6)	57.1 ± 7.57 (8)
Small Intestine	14.9 ± 3.00 (8)	11.9 ± 2.31 (12)	2.05 ± 0.85 (8)	0.68 ± 0.36 (12)[*]

Values are means ± SD (N). Metals were analyzed by flame AAS following wet digestion in HNO_3. Treatment means were compared by Student's t-test, * $P < 0.01$.

Changes in organ Cu content imply but do not prove that functional Cu deficiency exists. Therefore, to demonstrate that functional Cu deficiency exists and to investigate several related antioxidant enzymes, livers from -Cu mice were analyzed and compared to livers from +Cu mice (Table 6). Livers from -Cu mice were, on the average, 15% heavier than livers from +Cu mice. Three cuproproteins made by liver (CP, Cu,Zn-SOD, and cytochrome c oxidase) were analyzed. One protein, CP, is secreted to plasma and was measured there. Levels of CP activity in the -Cu mice were less than 2% of those from +Cu mice (Table 6) similar to other data (Table 4). Activity of both cytochrome c oxidase and Cu-Zn-SOD were lower in -Cu mice compared to +Cu mice, 48% and 79% of +Cu controls, respectively. This demonstrates that functional Cu deficiency exists in livers of -Cu mice.

Livers were also assayed for three non-cuproenzymes important in antioxidant metabolism. Liver GSH Px activity was not altered by dietary Cu deficiency in this model whereas liver GSH transferase was (Table 6). On the average, liver GSH transferase activity of -Cu mice was 76% of that of +Cu mice. Catalase activity was not different between +Cu and -Cu mice. Thus, for liver of -Cu mice the only reduction in an intracellular antioxidant enzyme was a modest drop in Cu,Zn-SOD.

Lymphoid organs were also analyzed from 7-week-old C57BL mice following Cu deficiency. Samples were taken randomly from many litters. Both absolute and relative weight of primary (thymus) and secondary (spleen) lymphoid tissues were altered by Cu deficiency (Table 7). In confirmation of previous work (Prohaska et al., 1983), the -Cu C57BL mice in these experiments demonstrated larger spleens and smaller thymuses. Both lymphoid organs, however, demonstrated, qualitatively and quantitatively, similar functional Cu deficiency as evidenced by significant reductions in Cu,Zn-SOD and cytochrome c oxidase activities compared to +Cu controls, for spleen 78% and 39% of +Cu and for thymus 74% and 38% of

Table 6. Liver Weight and Enzyme Activities of 7-week-old Male C57BL Mice Following Dietary Copper Deficiency.

Parameter	Cu-Adequate	Cu-Deficient
Liver Wt (g)	1.03 ± 0.14 (30)	1.19 ± 0.18 (30)[*]
Ceruloplasmin	47.7 ± 6.1 (13)	0.73 ± 1.02 (11)[*]
Cu,Zn-SOD	386 ± 76 (17)	306 ± 59 (18)[*]
Cyto Ox	1.22 ± 0.15 (10)	0.58 ± 0.27 (10)[*]
GSH-Px	2.86 ± 0.31 (9)	2.65 ± 0.32 (9)
GSH-T	0.95 ± 0.13 (9)	0.72 ± 0.08 (14)[*]
Catalase	623 ± 137 (14)	669 ± 151 (14)

Values are means ± SD (N). Enzyme activities, units/mg protein or units/L (ceruloplasmin), were determined spectrophotometrically. Means were compared by Student's t-test, * $P < 0.01$.

Table 7. Effect of Dietary Copper Deficiency on Lymphoid Tissues of 7-week-old Male C57BL Mice.[a]

	Spleen		Thymus[b]	
Parameters	+Cu	-Cu	+Cu	-Cu
Weight (mg)	64.2 ± 9.54 (10)	99.7 ±14.8 (10)[*]	44.3 ± 8.01 (10)	24.1 ± 2.78 (10)[*]
Wt/Body Wt (mg/g)	2.78 ± 0.33 (10)	4.66 ± 0.79 (10)[*]	1.94 ± 0.42 (10)	1.12 ± 0.13 (10)[*]
Cu (mg/g)	1.39 ± 0.50 (8)	0.37 ± 0.07 (7)[*]	0.94 ± 0.06 (4)	0.68 (1)
Fe (mg/g)	328 ±140 (8)	136 ±78.9 (8)[*]	17.6 ± 2.33 (4)	31.1 (1)
Cu,Zn-SOD (units/mg)	78.4 ± 4.64 (9)	61.3 ± 7.49 (7)[*]	68.1 ± 6.73 (9)	50.5 ± 8.39 (6)[*]
Cyto Ox (units/mg)	0.222 ± 0.026 (9)	0.086 ± 0.022 (7)[*]	0.250 ± 0.043 (9)	0.095 ± 0.034 (6)[*]

[a]Values are means ± SD (N). For each tissue, means were compared by Student's t-test, * $P < 0.01$
[b]Metal analyses of thymus tissue was based on pooling. For +Cu each pool contained 3 or 4 thymuses and for the -Cu pool 8 thymuses.

+Cu, respectively. As predicted from these cuproenzyme measurements the concentration of Cu was lower in both spleen and thymus of -Cu compared to +Cu mice (Table 7). Spleen Fe levels of -Cu mice were lower compared to +Cu mice in these samples.

In some experiments C57BL -Cu mice were repleted with Cu by switching a portion of the -Cu mice to the +Cu treatment for two weeks. The -Cu mice that were continued on the treatment still had lower thymus weights compared to +Cu controls, however, those that were repleted for 2 weeks had thymus weights equivalent to +Cu mice (Figure 2). Thus, the thymic hypoplasia of -Cu mice is reversible with Cu therapy.

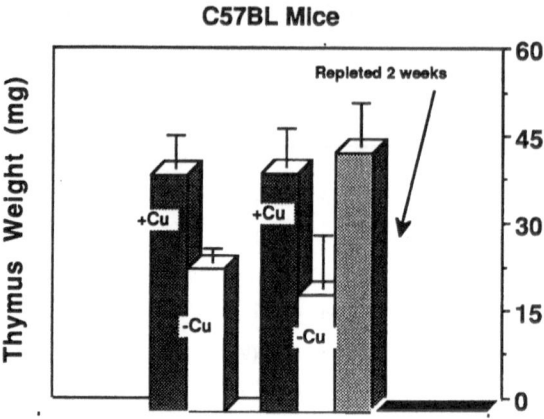

Figure 2. Thymus Weights Following Copper Deficiency and Repletion.

Immunological Changes Following Dietary Copper Deficiency

Following either seven or ten weeks of -Cu treatment male offspring were compared to +Cu offspring. The mice used for immunological assessment were first compared for Cu status and growth parameters as it was necessary to verify these measurements for individual correlation to immune status. As was observed previously, overall growth was not impaired by Cu deficiency (Table 8).

The 7-week old -Cu mice used for immunoassays were similar in their characteristics to the larger data sample of -Cu mice presented previously. For example, the deficit in thymus

Table 8. Effect of Copper Deficiency on Copper Status and Growth of Male C57BL Mice.

Parameter	Time on Diet (Weeks)			
	Seven		Ten	
	+Cu	-Cu	+Cu	-Cu
Body Wt (g)	23.2 ± 0.56	20.4 ± 1.68	25.0 ± 0.58	23.2 ± 0.24
Spleen Wt (mg)	74.5 ± 3.52	100 ± 5.61*	86.8 ± 3.09	85.8 ± 3.09
Thymus Wt (mg)	44.4 ± 3.83	16.0 ± 1.51*	42.0 ± 3.22	46.6 ± 2.92
Hematocrit (%)	52 ± 0.48	37 ± 4.45*	50 ± 0.63	52 ± 0.64
Liver Cu (μ/g)	5.24 ± 0.22	2.28 ± 0.25*	4.90 ± 0.04	3.88 ± 0.10*
Cu,Zn-SOD (U/mg)				
Spleen	37.8 ± 2.44	27.3 ± 1.96*	37.1 ± 1.39	42.0 ± 2.38
Thymus	51.2 ± 1.19	39.4 ± 1.43*	52.0 ± 3.03	45.2 ± 2.35

Values are means ± SEM (N = 4). For each age group, treatment means were compared by Student's t-test, * $P < 0.01$. Treatment was started at birth via dams.

weight and enlargement of spleen (Table 8) is similar to that seen in other mice (Table 7). The lower hematocrit and liver Cu in these -Cu mice (Table 8) is characteristic of C57BL 7-week old -Cu mice (Tables 4 and 5). The relative reduction in spleen and thymus Cu,Zn-SOD activities, 72% and 77% of +Cu values (Table 8), was also similar to other -Cu C57BL mice (Table 7).

In contrast, when the -Cu mice were kept on treatment for an additional 3 weeks their "Cu status" improved even though no supplement was provided. For example, the only significant difference between +Cu and -Cu mice when 10-weeks of age was a modest reduction in liver Cu (Table 8). This phenomena of improvement in Cu status when age or length of -Cu treatment was observed previously in C58 mice (Lukasewycz and Prohaska , 1982).

Splenocytes from the 7- and 10-week old mice were evaluated *in vitro* for immunological competence by measuring mitogen reactivity and ability to produce antibodies against sheep red blood cells (SRBC). In general, the -Cu 7-week-old mice demonstrated diminished responses whereas 10-week-old -Cu mice did not, corresponding nicely to their "Cu status." Splenocytes from 7-week-old -Cu mice demonstrated reduced proliferation to the T-cell mitogens Con A and PHA, as well as to the B-cell mitogen LPS. Response to the B-cell and T-cell stimulator PWM was equivocal (Table 9). The incorporation of ^{3}H-thymidine in the absence of mitogen was not statistically higher in cells from 7-week old -Cu mice compared to +Cu mice but was following 10 weeks of treatment (Table 9). The mitogen comparisons in cells from the 10-week-old mice were similar with the exception of a higher LPS response in -Cu compared to +Cu mice. The higher DNA synthesis in splenocytes in the absence of mitogen and reduced proliferation in the presence of mitogen has been a characteristic feature of -Cu C58 mice (Lukasewycz and Prohaska, 1983).

Table 9. Effect of Copper Deficiency on Mitogen Reactivity and Antibody Producing Cells from Spleens of Male C57BL Mice.[a]

Parameter	Time on Diet (Weeks)			
	Seven		Ten	
	+Cu	-Cu	+Cu	-Cu
Mitogen (CPM)				
None	2732 ± 460	3784 ± 542	4000 ± 187	6251 ± 537[*]
Con A	255817 ± 16724	176886 ± 4473[*]	248454 ± 3133	255090 ± 6531
PHA	75927 ± 7095	55546 ± 3210[†]	87540 ± 1078	84646 ± 1484
PWM	63594 ± 8807	58263 ± 4005	79981 ± 3345	90542 ± 5464
LPS	221753 ± 1379	171399 ± 2650[*]	172843 ± 3012	184465 ± 2323[†]
PFC[b]	446 ± 28.5	121 ± 17.7[*]	2732 ± 64.7	2581 ± 94.8
Viability (%)	64 ± 1.85	52 ± 1.32[*]	52 ± 0.75	52 ± 2.6

[a]Values are means ± SEM (N = 4). For each age group treatment means were compared by Student's *t*-test, * P < 0.01, † P < 0.05.
[b]PFC refers to hemolytic plaque-forming cells per culture, corrected for viability, following *in vitro* immunization with sheep red blood cells.

The same cells demonstrating a diminished mitogen response also produced fewer cells secreting IgM against SRBC *in vitro* (Table 9). In fact, a major reduction was evident (27% of +Cu). In contrast, the PFC response in splenocytes from -Cu 10-week old mice was normal. The survival (viability following four days of culture) of splenocytes from 7-week-old -Cu was lower than cells from +Cu mice (Table 9). Note, however, that the PFC data corrects for this and thus both antibody production and mitogen reactivity were significantly lower in C57BL mice that showed signs of Cu deficiency.

Several experiments were conducted to determine the immunoreactivity of thymocytes from 7-week-old -Cu mice compared to controls. Single cell suspensions of thymocytes were cultured with T-cell mitogens Con A and PHA. The level of DNA synthesis was determined by measuring ^{3}H-thymidine incorporation. There were no significant differences between cells from +Cu and -Cu mice for background radioactivity or response to either Con A or PHA (Table 10). Thus, despite the fewer number of thymocytes in -Cu mice the residual cells exhibit normal mitogen reactivity.

Table 10. Mitogen Reactivity of Thymocytes from 7-week-old C57BL Mice Following Dietary Copper Deficiency.

Mitogen	Cu-Adequate (5)	Cu-Deficient (5)
none (CPM)	273 ± 36.3	245 ± 56.3
Con A (SI)	253 ± 57.1	356 ± 102
PHA (SI)	24.2 ± 14.0	17.6 ± 6.0

Values are means ± SEM from separate experiments. Each experiment consisted of a pool of thymuses run in triplicate wells at several doses. Results presented are for Con A (0.5 µg/ml) and PHA (5 µg/ml). Means were equivalent by Student's t-test, P > 0.05.

Table 11. Effect of Dietary Copper Deficiency and Thioglycolate Injection on 7-week-old C57BL Mice.[a]

Parameter	Cu-Adequate (3)	Cu-Deficient (4)
Body Wt (g)	19.2 ± 0.3	17.3 ± 2.3
Spleen Wt (mg)	98.3 ± 10.1	155 ± 20[*]
Thymus Wt (mg)	54.2 ± 1.4	20.0 ± 7.0[*]
Hematocrit (%)	52.3 ± 1.7	23.8 ± 10.4[*]
Ceruloplasmin (U/L)[b]	22.4 ± 0.53	0.15 ± 0.11[*]
Cytochrome Ox (units/mg)		
Spleen	0.13 ± 0.01	0.05 ± 0.004[*]
Thymus	0.38 ± 0.10	0.24 ± 0.07
Macrophage	0.069 ± 0.020	0.019 ± .003[*]
Cu,Zn-SOD (units/mg)		
Spleen	52.5 ± 0.45	34.8 ± 5.0[*]
Thymus	59.6 ± 4.0	36.5 ± 1.9[*]
Macrophage	26.7 ± 4.2	13.1 ± 2.5[*]

[a]Values are means ± SD for male mice three days after receiving a 2.5 ml intraperitoneal injection of thioglycolate broth (3%). The elicited peritoneal macrophages were washed and assayed. Means were compared by Student's t-test, * P < 0.01.
[b]Assay was run at 30° and pH 5.0.

The third major cell type in the immunological triad of host defense is the macrophage. For these experiments 7-week-old +Cu and -Cu mice were studied following a sterile inflammatory response elicited by intraperitoneal injection of thioglycolate. The cells from the peritoneal cavity were freed of red cells by osmotic treatment and analyzed for Cu,Zn-SOD and cytochrome c oxidase activity (Table 11). For comparison and validity other parameters were also determined. The -Cu mice had features characteristic of other -Cu C57BL mice sampled previously including changes in lymphoid organ weight and enzyme activities as well as anemia and low plasma CP activity (Table 11). The deficit in cuproenzyme activities in macrophages from -Cu mice was even more pronounced than in splenocytes. For example, cytochrome c oxidase activity in macrophages from -Cu mice was 28% that of +Cu cells compared to 38% for splenocytes; for Cu,Zn-SOD the comparison was 49% for -Cu macrophages compared to 67% for splenocytes. Thus, the activated macrophages from -Cu mice exhibit functional Cu deficiency as severe as splenocytes. Immunological function of these macrophages was not performed in these studies.

Changes Following Genetic Copper Deficiency

Extensive previous studies comparing dietary to genetic copper deficiency in suckling mice indicated that both dietary deficient offspring (-Cu) and mice hemizygous for the brindled gene (Mo^{br}/y) exhibited signs characteristic of severe copper deficiency (Prohaska, 1983). Notably, both thymus and spleen were smaller in the 12-day-old -Cu and Mo^{br}/y offspring compared to controls. In these experiments thymus samples from several litters of mice were analyzed by flame AAS for total Cu and Fe content. Surprisingly, no significant reduction in thymus Cu content was observed in organs from either -Cu or Mo^{br}/y mice compared to their respective controls (Table 12). This result was quite unexpected since previous work had shown that Cu,Zn-SOD and cytochrome c oxidase activity was lower in thymus from similar mice (Prohaska, 1983). This paradox will require further research to resolve.

Table 12. Effect of Dietary and Genetic Copper Deficiency on Thymus Copper and Iron Levels in 12-day-old Mice.

Metal	+Cu	-Cu	+/y	Mo^{br}/y
Copper (µg/g)	0.85 ± 0.31 (5)	0.77 ± 0.27 (9)	0.87 ± 0.17 (9)	0.99 ± 0.46 (9)
Iron (µg/g)	7.80 ± 2.71 (5)	10.5 ± 1.79 (7)	8.55 ± 2.04 (8)	8.77 ± 2.39 (8)

Values are means ± SD for the numbers (in parentheses) of suckling male C57BL mice.

Some abnormal features of female heterozygotes carrying the brindled gene (+/Mo^{br}) are expressed due to random inactivation of the X-chromosome. Thus, it was of interest to ascertain whether immune function might be compromised in these mice. Splenocytes from adult C57BL female mice were cultured and tested for mitogen reactivity (Table 13). In comparison to controls (+/+), the reactivity of cells from brindled females (+/Mo^{br}) was equivalent for three mitogens tested (Con A, PHA and LPS). No difference in DNA synthesis in the absence of mitogen was noted between the two groups of mice. Thus, in this limited study it appears that the heterozygous brindled mice are not immunocompromised.

SUMMARY

Copper deficiency causes alterations in many biological systems because the lack of Cu alters the activity of ubiquitous cuproenzymes. The immune system is one such system that is altered by Cu deficiency. However, the immunological and biochemical mechanisms for these changes remain unknown.

Table 13. Mitogen Reactivity of Splenocytes from Adult Female Brindled Mice.

Mitogen	Control (+/+)		Brindled (Mo^{br}/+)	
none (CPM)	1193	± 492	2071	± 1448
Con A (SI)	42.8	± 9.95	38.6	± 11.1
PHA (SI)	17.8	± 7.38	11.6	± 3.05
LPS (SI)	33.4	± 3.35	23.8	± 11.7

Values are means ± SD for 6 adult C57BL mice of each genotype. Average CPM and stimulation indices (SI) were calculated from cells of individual mice run in triplicate. Only responses to optimal dose of mitogen are given. Responses at 2 other doses for each mitogen were equivalent between the two groups of mice, as were responses above, $P > 0.05$.

Immunological Background

In the past three years there have been a number of immunological studies published that have been confirmatory and complementary to initial studies conducted between 1981-1985 regarding copper and the immune system. Recent efforts are beginning to unravel the functions of copper in host defenses. Research in humans is limited due to the low frequency of Menkes' disease and the paucity of clinical documentation of nutritional copper deficiency in humans. Thus, research on domestic and laboratory animals will be important in delineating the copper-dependent mechanisms of impaired immune function. Studies with mice and rats have shown that copper is important for both antibody- and cell-mediated immunity and inflammatory responses.

We first showed an impairment in humoral immunity in -Cu mice by demonstrating a pronounced reduction in spleen antibody-producing cells when challenged with sheep red blood cells (SRBC) (Prohaska and Lukasewycz, 1981). We and others have confirmed this work with SRBC using both mice (Prohaska and Lukasewycz, 1989; Blakely and Hamilton, 1987) and rats (Vyas and Chandra, 1983; Failla et al., 1988). Another study in -Cu rats confirmed impaired antibody production to another T-dependent antigen, keyhole limpet hemocyanin (KLH) (Koller et al., 1987). Preliminary studies with Swiss albino mice (Lukasewycz et al., 1988) and Wistar rats (Eason et al., 1988) have shown an overall reduction in plasma IgG levels in Cu deficient animals. Thus specific as well as general antibody synthesis is impaired by dietary Cu deficiency.

We have also shown that cell-mediated immunity (CMI) is impaired in -Cu mice. In vivo mortality was high when we studied T-cell-dependent immunity to syngeneic malignant lymphocytes (Lukasewycz and Prohaska, 1982). Others have examined CMI in vivo using delayed-type hypersensitivity (DTH) with mixed outcomes. One study reported enhanced DTH in -Cu mice (Jones, 1984) while studies in -Cu rats have reported normal (Koller et al., 1987) or suppressed DTH (Kishore et al., 1984).

To complement these studies and to explain the in vivo impairment in CMI and antibody production several *in vitro* procedures have been used. We examined previously -Cu C58 mice splenic lymphocyte responses to stimulation by several mitogens and found reduced proliferation to both T- and B- specific probes (Lukasewycz and Prohaska, 1983; Lukasewycz et al., 1985). Kramer and coworkers have confirmed that splenocytes from -Cu rats have reduced proliferation to Con A (a T-cell mitogen) and that adherent cells (presumably macrophages) may be involved (Davis et al., 1987). Recently, we (Prohaska and Lukasewycz, 1989) and others (Blakely and Hamilton, 1987) have found overall proliferation to be normal in cultures of residual splenocytes from -Cu outbred Swiss albino mice. We have also examined allogeneic mixed lymphocyte reactions (Lukasewycz et al., 1987) and shown that splenocytes from -Cu mice are poor responders and weak stimulators in MLR. One report showed that -Cu rats have impaired natural killer (NK) cell activity (Koller et al., 1987). Thus, there is good evidence from a number of labs that copper nutriture is critical for both antibody- and cell-mediated immunity. What, then, are the immune mechanisms involved?

We have shown that chronic copper deficiency alters lymphocyte distribution in spleens of -Cu C58 mice, resulting in a significant increase in B-cells and a relative decrease in T-cells, especially within the Lyt 1 (helper) subset (Lukasewycz et al., 1985). Mulhern and Koller (1988) working with -Cu C57BL mice have confirmed these observations. This shift in lymphocyte distribution may influence the immune response of -Cu mice.

Perhaps it is not only distribution but also the functional properties of cells that are important. Lymphokines produced by T-cells are crucial in the coordinate regulation of immune and inflammatory responses (Miyajima et al., 1988). A key factor in sorting out the complex observations following copper deficiency (including anemia and neutropenia) will be a determination of lymphokine and monokine metabolism. Studies by others using *in vitro* procedures with media that is "copper-deficient" have shown that splenocytes in "-Cu" cultures have an impaired production of both the monocyte-derived factor, interleukin-1 (IL-1) (Flynn et al., 1984), and the T-cell-replacing factor (now known as IL-5) (Flynn and Yen, 1981) when studied in allogeneic lymphocyte cultures. A study in -Cu rats indicated that serum thymic factor levels were low in -Cu animals (Vyas and Chandra, 1983). We are excited about our recent preliminary studies that show reduced IL-2 production (a lymphokine from type I helper T-cells, T_H1) (Prohaska et al., 1988a). It will be important to determine if production of other T_H1 specific lymphokines such as interferon-γ (INF-γ) are altered. Mulhern and coworkers have shown that bone marrow from -Cu NZB (Mulhern et al., 1987) and C57BL (Mulhern and Koeller, 1988) mice induced colony formation in spleens of irradiated mice to much greater degree compared to +Cu controls indicating a stimulation in erythropoiesis. We (Lukasewycz and Prohaska, 1983; Lukasewycz et al., 1985) and others (Mulhern and Koeller, 1988) have reported that -Cu splenocytes exhibit greater cell cycling (DNA synthesis) in the "unstimulated" state. Recently we have extended these observations on DNA synthesis in an additional mouse strain (Prohaska and Lukasewycz, 1989). Perhaps this indicates an imbalance in lymphokine production in Cu deficiency. These observations will be important to consider when investigating the role of copper in lymphocyte and granulocyte differentiation.

In studies with -Cu cattle (Jones and Suttle, 1981; Boyne and Arthur, 1981 & 1986) and sheep (Jones and Suttle, 1981) it was shown that neutrophil function was altered. These were studies with cells from peripheral blood. The cells from -Cu animals demonstrated normal capacities of phagocytosis in all but one case (Boyne and Arthur, 1986) but impaired microbicidal activity against *Candida albicans* . Hypocupremic infants also appear to have impaired phagocytic function (Heresi et al., 1985).

Surprisingly little research has been done with macrophages from Cu deficient animals. Many of the above observations in -Cu rodents could be explained by altered macrophage function since antigen presentation and interleukin-1 (IL-1) (a monokine produced by macrophages) are both needed to activate antibody producing B-cells (Hoffman et al., 1985). IL-1 is also necessary for T-lymphocyte activation by mitogens (Rosenstreich et al., 1976). A provocative paper on zinc deficiency has suggested, in fact, that the lymphocyte proliferation differences in -Zn mice were due to differences in macrophages (James et al., 1987). Our preliminary work with splenic and peritoneal macrophages indicate -Cu mice produce greater amounts of IL-1 (Lukasewycz and Prohaska, 1989). It would thus appear that both lymphocyte and macrophage function are sensitive to changes in copper status but in different ways.

Certainly long range investigations of copper functions in the immune system should include lymphocytes, monocytes and granulocytes. The alterations in immune function in copper-deficient rodents are correlated with "Cu status" and not necessarily with the associated anemia (Kramer et al., 1988). Thus, it will be important to establish the copper-dependent mechanisms of altered immunity that occur when dietary and genetic copper deficiency exist.

Biochemical Background

Biochemical and morphological information on spleen and thymus of -Cu rodents has been published. We have produced dietary copper deficiency in several strains of mice including C58, C57BL (in the current studies), albino, and Swiss albino with comparable characteristics. The -Cu mice develop anemia and have low activity of ceruloplasmin (EC 1.16.3.1). They have enlarged spleens and small thymus glands. Moreover, both organs demonstrate functional copper deficiency, having decreased levels of two cuproenzymes, copper-zinc superoxide dismutase (Cu,Zn-SOD, EC 1.15.1.1) and cytochrome c oxidase (EC 1.9.3.1) (Prohaska et al., 1983). Others have confirmed the changes in lymphoid organ size in -Cu mice (Mulhern and Koeller, 1988). A small thymus is also observed in -Cu rats (Faillia et al., 1988; Koeller et al., 1987). Likewise lymphoid tissues of -Cu rats also have lower levels of Cu (Faillia et al., 1988) and cytochrome c oxidase activity (Davis et al., 1987). Results from electron microscope studies (Prohaska et al., 1983) indicated necrotic changes in both spleen and thymus from -Cu mice involving mitochondria and nuclei. Peritoneal macrophages from --- Cu mice exhibit lower activity of cytochrome c oxidase and Cu,Zn-SOD, as indicated in the current experiments and show changes in morphology (unpublished). Thus, chronic dietary copper deficiency results in a functional copper deficiency and altered morphology in lymphocytes and macrophages. Several alternative biochemical hypotheses can be stated which could explain the impaired immune response observed in experimental rodents.

Hypothesis 1: Impaired immunity is due to changes in plasma membranes. Our MLR study,which showed that inactivated spenocytes from -Cu mice were poor stimulators, suggested that chronic copper deficiency has altered the lymphocyte membrane (Lukasewycz et al., 1987). This might have been due to altered expression of Ia antigen (a class II surface glycoprotein encoded by the major histocompatibility complex) or a non-specific effect on membrane structure. We tested the latter possibility by purifying and characterizing lymphocyte plasma membranes (Korte and Prohaska, 1987). We found, especially in splenocyte preparations, altered fatty acid composition. Polypeptide profiles as viewed by SDS-PAGE techniques were similar between samples from +Cu and -Cu mice. We plan to measure membrane fluidity as recent work by others suggests that copper deficiency does alter membrane properties. Liver plasma membranes from -Cu rats showed depressed fluidity (Lei et al., 1988). The protein (Johnson and Kramer, 1987) and lipid (Jain and Williams, 1988) composition of erythrocyte membranes from -Cu rats has been shown to be altered. This resulted in increased erythrocyte viscosity (Jain and Williams, 1988).

Alterations in the membrane may be caused by defective synthesis or increased catabolism. It is known that -Cu rats exhibit hypercholesterolemia (Allen and Klevay, 1978). This may influence membrane cholesterol levels of immunocompetent cells but this has not been determined. It is known that an impaired immune response occurs when membrane cholesterol levels are altered (Duwe et al., 1981). Changes in membrane phospholipid fatty acid content that we measured, including polyunsaturated fatty acids (PUFA), do not support the notion of increased lipid peroxidation (Korte and Prohaska, 1987) even though spleens of -Cu mice have low activities of both the antioxidant enzyme Cu,Zn-SOD (Prohaska et al., 1983) and the antioxidant L-ascorbic acid (Prohaska and Cox, 1983; Prohaska et al., 1984). Furthermore, others have recently shown that -Cu rats are susceptible to lipid peroxidation (Lawrence and Jenkinson, 1987).

The biosynthesis of lymphocyte plasma membranes requires energy. It is therefore conceivable that decreased activity of CO may limit ATP flux and thus alter membrane protein and phospholipid content. We have shown that the hearts of -Cu mice have reduced ATP levels and adenylate energy charge (Rusinko and Prohaska, 1985). Changes in the protein and lipid content of the plasma membrane might also impair the immunoreactivity of lymphocytes and macrophages.

Hypothesis 2: Altered immunity is due to changes in the intracellular redox status and energy metabolism. The rationale for increased lipid peroxidation as described above would not only alter membrane composition but could also alter the redox status such that changes in glutathione (GSH/GSSG) would occur. This could influence cell proliferation (Noelle and Lawrence, 1981) and antibody production (Thomas and Holt, 1978). Recently, altered GSH levels have been reported in -Cu rats (Allen et al., 1988). Moreover, altered pyridine nucleotide ratios have been demonstrated in brain of -Cu rats (Prohaska and Wells, 1975). It was shown that an elevation in lactate/pyruvate and α-glycerol phosphate/dihydroxyacetone phosphate existed suggesting an elevation in NADH/NAD+.

Hypothesis 3: Altered immunity is due to changes in neuroendocrine function. There exists a complex interaction between the immune, nervous, and reproductive systems. This hypothalamic-pituitary-gonadal-thymic axis is regulated by a multifactorial interdependent process (Grossman, 1985). Perhaps no other area of copper research is as exciting and potentially important. Some of the unexplained observations associated with copper deficiency may lie in this complex interplay. Many key molecules are involved and several may depend on copper.

For example, within the nervous system norepinephrine (NE) synthesis depends on dopamine-β–monooxygenase (DBM). It is known that NE turnover is decreased during the immune response (Besedovsky et al., 1983). Preliminary work in heart has shown that NE turnover is elevated by Cu deficiency (Gross and Prohaska, 1989). Furthermore, we have shown that spleen levels of NE are lower in -Cu mice (Prohaska and DeLuca, 1988). Thus, NE metabolism, known to be altered by Cu deficiency and to be important in such processes as regulation of cytosolic calcium (Erickson et al., 1987), may be a key component in explaining altered immune function.

Another enzyme, peptidyl glycine α–amidating monooxygenase (PAM), may also be important. This enzyme, like DBM, requires copper and ascorbate as cofactors (Eipper et al., 1983). It is responsible for posttranslationally modifying a large number of neuroactive peptides. One such protein is vasoactive intestinal peptide (VIP). VIP couples with its lymphocyte receptor to activate adenylate cyclase which leads to a cAMP-dependent cascade (O'Dorisio et al., 1985). VIP has a profound influence on the immune system and its synthesis is dependent on a cuproprotein. It is not known if Cu deficiency alters levels of VIP and other neuroactive peptides dependent on PAM for synthesis.

Another site in the nervous system that may respond to changes in dietary copper is the hypothalamus. From the work of Barnea and colleagues (Barnea et al., 1988) we know that the release of luteinizing hormone releasing hormone (LHRH) from the median eminence area is greatly amplified in the presence of copper and prostaglandin E_2 (PGE_2). Release of LH from the pituitary has a major effect on gonadal hormone release which in turn affects the thymus (Grossman, 1985).

In mice, steroid hormones such as corticosterone, estrogen, and testosterone influence the immune system and thymus size. Preliminary studies in our lab indicate that male -Cu mice have low plasma testosterone (Korte et al., 1988). Others studying -Cu rats have reported low plasma testosterone and estradiol levels (Farquharson and Robins, 1988). Elevated corticosterone is partly responsible for the small thymus observed in zinc deficiency (Depasquale-Jardieu and Fraker, 1980). This appears not to be the case in nutritional copper deficiency (Prohaska et al., 1983).

There are other neuroendocrine factors that might be altered by Cu deficiency. The enkephalins, for example, may be one such factor. The factors discussed here were meant to serve as reasonable possibilities for future research. A reduction in the pool size of NE, VIP, LH, thymulin, and gonadal hormones in copper deficiency would influence immunoregulation.

Hypothesis 4: Altered immunity is due to changes in some unknown copper-dependent factor. Very little information is available concerning copper metabolism in lymphocytes. We do know lymphocytes contain Cu,Zn-SOD and cytochrome c oxidase, two copper-dependent factors. Lymphocytes possess ceruloplasmin receptors and thus have a specific mechanism for copper uptake from blood. In view of the importance of copper in immune function, it seems timely to begin a thorough study into the levels, distribution, and turnover of copper in lymphoid organs. Lymphocytes may contain unique enzymes dependent on copper for activity. In a deficient state, changes in these enzymes may influence the immune system more directly than the changes mentioned above dealing with ceruloplasmin, Cu,Zn-SOD, or cytochrome c oxidase. One such candidate enzyme is the sulfhydryl oxidase located in the plasma membrane of lymphocytes (Roth and Koshland, 1981). This enzyme is important in pentamer IgM formation and thus B-cell differentiation. In addition, copper seems to influence eicosanoid production. Changes in levels of PGE_2, for example, would impact host defenses. Recently, Allen and coworkers have demonstrated in -Cu rats that both prostacyclin and prostaglandin E_2 and $F2_\alpha$ production were decreased (Mitchell et al., 1988; Lampi et al., 1988). However, resident peritoneal macrophages from -Cu rats produced normal quantities of PGE_2 when stimulated with LPS (Koeller et al., 1987).

Pathogenesis of immunologic dysfunction in copper deficiency may include several factors. Evaluation of these possibilities in lymphocytes and macrophages will be necessary to characterize fully the consequences of copper deficiency to the developing immune system.

ACKNOWLEDGEMENTS

Supported in part by NIH grant HD 20975. Technical assistance of William Bailey, Karen Kolquist, and Joseph Korte is appreciated.

REFERENCES

Allen, K. G. D., and Klevay, L. M., 1978, Cholesterolemia and cardiovascular abnormalities in rats caused by copper deficiency, **Atherosclerosis**, 29:81-93.

Allen, K. G. D., Arthur, J. R., Morrice, P. C., Nicol, F., and Mills, C. F., 1988, Copper deficiency and tissue glutathione concentration in the rat (42634), **Proc. Soc. Exp. Biol. Med.**, 187:38-43.

Barnea, A., Cho., G., and Hartter, D. E., 1988, A correlation between the ligand specificity for [67]copper uptake and for copper-prostaglandin E_2 stimulation of the release of gonadotropin-releasing hormone from median eminence explants, **Endocrinology**, 122:1505-1510.

Bell, J. G., Keen, C. L., and Lönnerdal, B., 1987, Effect of infant cereals on zinc and copper absorption during weaning, **Am. J. Dis. Child.**, 141:1128-1132.

Besedovsky, H., Del Rey, A., Sorkin, E., Da Prada, M., Burri, R., and Honegger, C., 1983, The immune response evokes changes in brain noradrenergic neurons, **Science**, 221:564-566.

Blakley, B. R., and Hamilton, D. L., 1987, The effect of copper deficiency on the immune response in mice, **Drug-Nutr. Interactions**, 5:103-111.

Boyne, R., and Arthur, J. R., 1981, Effects of selenium and copper deficiency on neutrophil function in cattle, **J. Comp. Pathol.**, 91:271-276.

Boyne, R., and Arthur, J. R., 1986, Effects of molybdenum or iron induced copper deficiency on the viability and function of neutrophils from cattle, **Res. Vet. Sci.**, 41:417-419.

Castillo-Duran, C., Fisberg, M., Valenzuela, A., Egaana, J. I., and Uauy, R., 1983, Controlled trial of copper supplementation during the recovery from marasmus, **Am. J. Clin. Nutr.**, 37: 898-903.

Danks, D. M., Campbell, P. E., Stevens, B. J., Mayne, V., and Cartwright, E., 1972, Menkes's kinky hair syndrome, **Pediatrics**, 50:188-201.

Davis, M. A., Johnson, W. T., Briske-Anderson, M., and Kramer, T. R., 1987, Lymphoid cell functions during copper deficiency, **Nutr. Res.**, 7:211-222.

DePasquale-Jardieu, P., and Fraker, P. J., 1980, Further characterization of the role of corticosterone in the loss of humoral immunity in zinc-deficient A/J mice as determined by adrenalectomy, **J. Immunol.**, 124:2650-2655.

Duwe, A. K., Fitch, M., and Ostwald, R., 1981, Effects of dietary cholesterol on antibody-dependent phagocytosis and cell-mediated lysis in guinea pigs, **J. Nutr.**, 111:1672-1680.

Eason, S., Carville, D., Strain, J. J., and Hannigan, B. M., 1988, The influence of dietary carbohydrate on antibody-mediated immunity in copper deficiency, **Biochem. Soc. Trans.**, 16:54-55.

Eipper B. A., Mains, R. E., and Glembotski, C. C., 1983, Identification in pituitary tissue of a peptide α-amidation activity that acts on glycine-extended peptides and requires molecular oxygen, copper, and ascorbic acid, **Proc. Natl. Acad. Sci. USA**, 80:5144-5148.

Erickson, R. R., Prasad, J. S., and Holtzman, J. L., 1987, The role of NADPH- and reduced glutathione-dependent enzymes in the norepinephrine modulation of the ATP-dependent, hepatic microsomal calcium pump: a new pathway for the noradrenergic regulation of cytosolic calcium in the hepatocyte, **J. Pharmacol. Exp. Ther.**, 242:472-477.

Failla, M. L., Babu, U., and Seidel, K. E., 1988, Use of immunoresponsiveness to demonstrate that the dietary requirement for copper in young rats is greater with dietary fructose than dietary starch, **J. Nutr.**, 118:487-496.

Farquharson, C., and Robins, S. P., 1988, Female rats are susceptible to cardiac hypertrophy induced by copper deficiency: the lack of influence of estrogen and testosterone, **Proc. Soc. Exp. Biol. Med.**, 188:272-281.

Flynn, A., Loftus, M. A., and Finke, J. H., 1984, Production of interleukin-1 and interleukin-2 in allogeneic mixed lymphocyte cultures under copper, magnesium and zinc deficient conditions, **Nutr. Res.**, 4:673-679.

Flynn, A., and Yen, B. R., 1981, Mineral deficiency effects on the generation of cytotoxic T-cells and T-helper cell factors *in vitro*, **J. Nutr.**, 111:907-913.

Gibson, R. S., 1985, Dietary intakes of trace elements in young children, **Food Nutr. News**, 57: 21-24.

Gross, A.M., and Prohaska, J.R., 1989, Copper-deficient mice have higher cardiac norepinephrine turnover, **Fed. Proc.** 48:in press.

Grossman, C. J., 1985, Interactions between the gonadal steroids and the immune sytstem, **Science**, 227:257-261.

Habig, W. H., Pabst, M. J., and Jakoby, W. B., 1974, Glutathione S-transferases. The first enzymatic step in mercapturic acid formation, **J. Biol. Chem.**, 249:7130-7139.

Haschke, F., Ziegler, E. E., Edwards, B. B., and Fomon, S. J., 1986, Effect of iron fortification of infant formula on trace mineral absorption, **J. Ped. Gastroenterol. Nutr.**, 5:768-773.

Heresi, G., Castillo-Durán, C., Muñoz, C., Arévalo, M., and Schlesinger, L., 1985, Phagocytosis and immunoglobulin levels in hypocupremic infants, **Nutr. Res.**, 5:1327-1334.

Hoffmann, M. K., Mizel, S. B., and Hirst, J. A., 1984, IL 1 requirement for B cell activation revealed by use of adult serum, **J. Immunol.**, 133:2566-2568.

Jain, S. K., and Williams, D. M., 1988, Copper deficiency anemia: altered red blood cell lipids and viscosity in rats, **Am. J. Clin. Nutr.**, 48:637-640.

James, S. J., Swendseid, M., and Makinodan, T., 1987, Macrophage-mediated depression of T-cell proliferation in zinc-deficient mice, **J. Nutr.**, 117:1982-1988.

Johnson, W. T., and Kramer, T. R., 1987, Effect of copper deficiency on erythrocyte membrane proteins in rats, **J. Nutr.**, 117:1085-1090.

Jones, D. G., 1984, Effects of dietary copper depletion on acute and delayed inflammatory responses in mice, **Res. Vet. Sci.**, 37:205-210.

Jones, D. G., and Suttle, N. F., 1981, Some effects of copper deficiency on leucocyte function in sheep and cattle, **Res. Vet. Science**, 31:151-156.

Jones, D. G., and Suttle, N. F., 1983, The effect of copper dieficiency on the resistance of mice to infection with *Pasteurella haemolytica*, **J. Comp. Pathol.**, 93:143-149.

Kishore, V., Latman, N., Roberts, D. W., Barnett, J. B., and Sorenson, J. R. J., 1984, Effect of nutritional copper deficiency on adjuvant arthritis and immunocompetence in the rat, **Agents and Actions**, 14:274-282.

Koller, L. D., Mulhern, S. A., Frankel, N. C., Steven, M. G., and Williams, J. R., 1987, Immune dysfunction in rats fed a diet deficient in copper, **Am. J. Clin. Nutr.**, 45:997-1006.

Korte, J.J., Bailey, W.R., and Prohaska, J.R., 1988, Copper deficiency impairs development of murine hepatic glutathione transferase MII, **Fed. Proc.**, 47: A1105.

Korte, J. J., and Prohaska, J. R., 1987, Dietary copper deficiency alters protein and lipid composition of murine lymphocyte plasma membranes, **J. Nutr.**, 117:1076-1084.

Kramer, T. R., Johnson, W. T., and Briske-Anderson, M., 1988, Influence of iron and the sex of rats on hematological, biochemical and immunological changes during copper deficiency, **J. Nutr.**, 118:214-221.

Lampi, K. J., Mathias, M. M., Rengers, B. D., and Allen, K. G. D., 1988, Dietary copper and copper dependent superoxide dismutase in hepatic prostaglandin synthesis by rat liver homogenates, **Nutr. Res.**, 8:1191-1202.

Lawrence, R. A., and Jenkinson, S. G., 1987, Effects of copper deficiency on carbon tetrachloride-induced lipid peroxidation, **J. Lab. Clin. Med.**, 109:134-140.

Lei, K. Y., Rosenstein, F., Shi, F., Hassel, C. A., Carr, T. P., and Zhang, J., 1988, Alterations in lipid composition and fluidity of liver plasma membranes in copper-deficient rats, **Proc. Soc. Exp. Biol. Med.**, 188:335-341.

Lukasewycz, O. A., Kolquist, K. L., and Prohaska, J. R., 1987, Splenocytes from copper-deficient mice are low responders and weak stimulators in mixed lymphocyte reactions, **Nutr. Res.**, 7:43-52.

Lukasewycz, O. A., Kolquist, K. L., and Prohaska, J. R., 1988, Modulation in immunoglobulin (Ig) isotype production in copper-deficient mice, **FASEB J.**, 2:A436.

Lukasewycz, O. A., and Prohaska, J. R., 1982, Immunization against transplantable leukemia impaired in copper deficient mice, **J. Natl. Cancer Inst.**, 69:489-493.

Lukasewycz, O. A., and Prohaska, J. R., 1983, Lymphocytes from copper-deficient mice exhibit decreased mitogen reactivity, **Nutr. Res.**, 3:335-341.

Lukasewycz, O. A., and Prohaska, J. R., 1989, Increased interleukin-1 (IL-1) and decreased interleukin-2 (IL-2) production in copper-deficient mice, **FASEB J.**, 3:A665.

Lukasewycz, O. A., Prohaska, J. R., Meyer, S. G., Schmidtke, J. R., Hatfield, S. M., and Marder, P., 1985, Alterations in lymphocyte subpopulations in copper-deficient mice, **Infect. Immun.**, 48:644-647.

Markwell, M. A. K., Haas, S. M., Bieber, L. L., and Tolbert, N. E., 1978, Modification of the Lowry procedure to simplify protein determination in membrane and lipoprotein samples, **Anal. Biochem.**, 87: 206-210.

Mishell, R. I., and Dutton R. W., 1967, Immunization of dissociated spleen cell cultures from normal mice, **J. Exp. Med.**, 126:423-442.

Mitchell, L. L., Allen, K. G. D., and Mathias, M. M., 1988, Copper deficiency depresses rat aortae superoxide dismutase activity and prostacyclin synthesis, **Prostaglandins**, 35:977-986.

Miyajima, A., Miyatake, S., Schreurs, J., De Vries, J., Arai, N., Yokota, T., and Arai, E.-I., 1988, Coordinate regulation of immune and inflammatory responses by T cell-derived lymphokines, **FASEB J.**, 2:2462-2473.

Mosmann, T. R., Cherwinski, H., Bond, M. W., Giedlin, M. A., and Coffman, R. L., 1986, Two types of murine helper T cell clone. I. Definition according to profiles of lymphokine activities and secreted proteins, **J. Immunol.**, 136:2348-2357.

Mulhern, S. A., and Koller, L. D., 1988, Severe or marginal copper deficiency results in a graded reduction in immune status in mice, **J. Nutr.**, 118:1041-1047.

Mulhern, S. A., Raveche, E. S., Smith, H. R., and Lal, R. B., 1987, Dietary copper deficiency and autoimmunity in the NZB mouse, **Am. J. Clin. Nutr.**, 46:1035-1039.

Newberne, P. M., Hunt, C. E., and Young, V. R., 1968, The role of diet and the reticuloendothelial system in the response of rats to *Samonella typhimurium* infection, **Br. J. Exp. Pathol.**, 49:448-457.

Noelle, R. J., and Lawrence, D. A., 1981, Determination of glutathione in lymphocytes and possible association of redox state and proliferative capacity of lymphocytes, **Biochem. J.**, 198:571-579.

O'Dorisio, M. S., Wood, C. L., and O'Dorisio, T. M., 1985, Vasoactive intestinal peptide and neuropeptide modulation of the immune response, **J. Immunol.**, 135:792s-796s.

Oppenheimer, S. M., Hoffbrand, B. I., Dormandy, T. L., Parker, N., and Wickens, D. G., 1987, Macrocytic anaemia due to copper deficiency in a patient with late onset hypogammaglobulinaemia, **Postgrad. Med. J.**, 63:205-207.

Pletcher, J. M., and Banting, L. F., 1983, Copper deficiency in piglets characterized by spongy myelopathy and degenerative lesions in the great blood vessels, **J. So.Af. Vet Assoc.**, 54:43-46.

Prohaska, J. R., 1983, Changes in tissue growth, concentrations of copper, iron, cytochrome oxidase and superoxide dismutase subsequent to dietary or genetic copper deficiency in mice, **J. Nutr.**, 113:2048-2058.

Prohaska, J. R., 1988, Biochemical functions of copper in animals, *in:* **"Essential and Toxic Elements in Human Health and Disease,"** A. S. Prasad, ed., Alan R. Liss, Inc., New York, NY, pp.105-124.

Prohaska, J. R., and Cox, D. A., 1983, Decreased brain ascorbate levels in copper-deficient mice and in brindled mice, **J. Nutr.**, 113:2623-2629.

Prohaska, J. R., Cox, D. A., and Bailey, W. R., 1984, Ascorbic acid synthesis and concentrations in organs of copper-deficient and brindled mice, **Biol. Trace Elem. Res.**, 6:441-453.

Prohaska, J. R., and DeLuca, K. L., 1988, Norepinephrine and dopamine distribution in copper-deficient mice, *in:* **"Trace Elements in Man and Animals 6"**, L. S. Hurley, C. L. Keen, B. Lonnerdal, and R. B. Rucker, eds., Plenum Press, New York, NY, pp. 109-111.

Prohaska, J. R., Downing, S. W., and Lukasewycz, O. A., 1983, Chronic dietary copper deficiency alters biochemical and morphological properties of mouse lymphoid tissues, **J. Nutr.**, 113:1583-1590.

Prohaska, J. R., and Lukasewycz, O. A., 1981, Copper deficiency suppresses the immune response of mice, **Science**, 213:559-561.

Prohaska, J. R., and Lukasewycz, O. A., 1989, Biochemical and immunological changes in mice following postweaning copper deficiency, **Biol. Trace Elem. Res.**, in press.

Prohaska, J.R., Solem, L.E., and Lukasewycz, O.A., 1988a Variation in interleukin-2 (IL-2) production by copper-deficient mice. **FASEB J.**, 2: A436.

Prohaska, J. R., and Wells, W. W., 1975, Copper deficiency in the developing rat brain: evidence for abnormal mitochondria, **J. Neurochem.**, 25:221-228.

Prohaska, J. R., Wittmers, L. E., and Haller, E. W., 1988b, Influence of genetic obesity, food intake, and adrenalectomy in mice on selected trace element-dependent protective enzymes, **J. Nutr.**, 118:739-746.

Rosenstreich, D. L., Farrar, J. J., and Dougherty, S., 1976, Absolute macrophage dependency of T lymphocyte activation by mitogens, **J. Immunol.**, 116:131-139.

Roth, R. A., and Koshland, M. E., 1981, Identification of a lymphocyte enzyme that catalyzes pentamer immunoglobulin M assembly, **J. Biol. Chem.**, 256:4633-46539.

Rusinko, N., and Prohaska, J. R., 1985, Adenine nucleotide and lactate levels in organs from copper-deficient mice and brindled mice, **J. Nutr.**, 115:936-943.

Sandstead, H. H., 1982, Copper bioavailability and requirements, **Am. J. Clin. Nutr.**, 35:809-814.

Simon, S. R., Branda, R. F., Tindle, B. H., and Burns, S. L., 1988, Copper deficiency and sideroblastic anemia associated with zinc ingestion, **Am. J. Hematol.**, 28:181-183.

Thomas, W. R., and Holt, P. G., 1978, Vitamin C and immunity: an assessment of the evidence, **Clin Exp. Immunol.**, 32:370-379.

Vyas, E., and Chandra, R. K., 1983, Thymic factor activity, lymphocyte stimulation response and antibody producing cells in copper deficiency, **Nutr. Res.**, 3:343-349.

SELENIUM AND GLUTATHIONE PEROXIDASE: ESSENTIAL NUTRIENT AND ANTIOXIDANT COMPONENT OF THE IMMUNE SYSTEM

Julian E. Spallholz

Texas Tech University
Institute for Nutritional Sciences and
Center for Food and Nutrition
Lubbock, TX 79409

INTRODUCTION

Dioxygen (O_2) is required by most plants and animals (aerobes) for the oxidation of carbohydrates (glucose principally), protein (amino acids) and lipids (fatty acids) which serves as the terminal acceptor in mitochondria of both hydrogen and electrons in the formation of metabolic water. For aerobic metabolism dioxygen is essential. Other forms of life, the anaerobes, are viable only in the absence of dioxygen, for dioxygen is toxic to anaerobic organisms. Aerobic life has thus evolved in a dioxygen environment having developed the necessary cellular defenses to thwart dioxygen toxicity.

Dioxygen toxicity arises from the thermodynamically favored reactions of O_2 and superoxide (O_2-) with biological molecules, particularly those exhibiting unsaturation. Fortunately, however, such reactions of O_2 and O_2- with biological molecules are slow and often free radicals (Table 1) react with other free radicals in annihilation reactions (Allen, 1986). When free radicals do not undergo annhilation reactions biological damage occurs and it is the result of uncontrollable oxidation by free radical chain reactions (Figure 1). Aerobic cells and higher orders of living plants, animals and humans have developed antioxidant defense mechanisms to prevent sustained free radical chain reactions. For animals and humans, free radical defense mechanisms include the natural antioxidants consumed in foods, antioxidants added to foods, small organic antioxidants synthesized by cells, and enzymes, many which contain minerals and are synthesized *in vivo* (Table 2).

Table 1. Free Radicals Produced by the Reduction of Dioxygen, by Ionizing Radiation, by Reactive Metals, Enzymes and Other Endogenous and Environmental Initiators.

$\cdot O_2-$	Superoxide
HO_2-	Superoxide Conjugate Acid
$^1O_2$	Singlet Oxygen
$\cdot OH$	Hydroxyl Radical
$R\cdot$	Organic Free Radical
$ROO\cdot$	Peroxy Free Radical

In this American Chemical Society Symposium on **Antioxidant Nutrients and the Immune Response,** many of the natural endogenous and exogenous antioxidants and minerals contained within the antioxidant enzymes were discussed in light of their known effects on the immune system. This paper's focus is specifically on the biological essentiality of selenium (Se), its inclusion in the antioxidant glutathione peroxidases and the role(s) of selenium and/or glutathione peroxidase in tissues and cells of the reticulo- endothelial (immune) system.

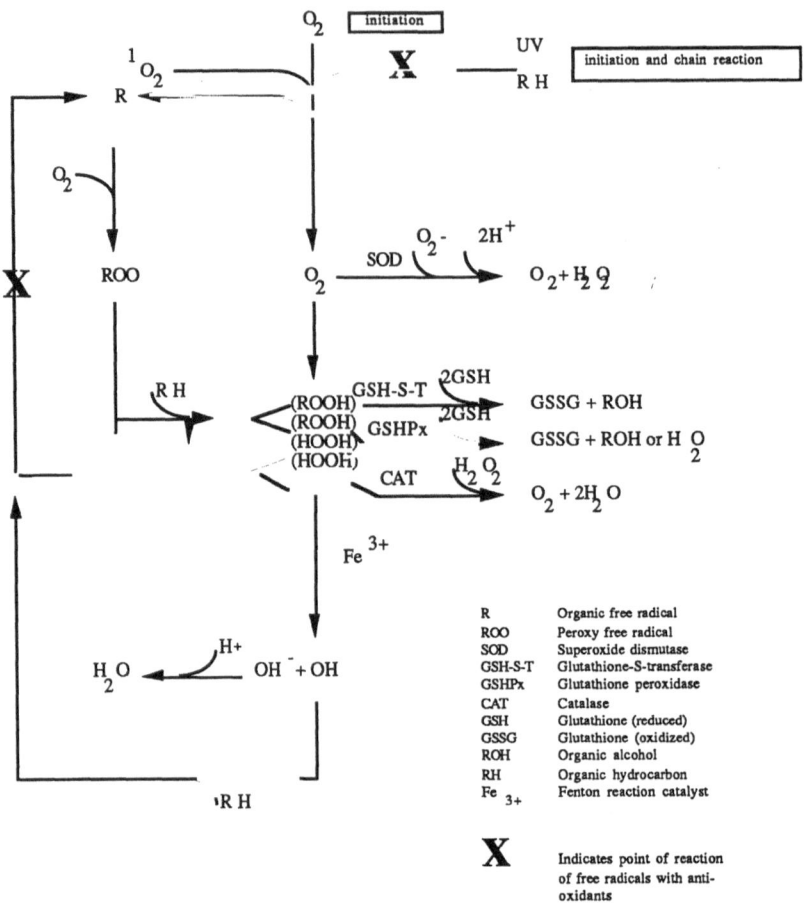

Figure 1. Free-radical chain reactions and their prevention. Modified from Spallholz, (1989).

Table 2. Exogenous and Endogenous Antioxidants Present in Many Aerobic Cells

Exogenous Antioxidants of Humans	Endogenous Antioxidants Synthesized *in Vivo*
Natural antioxidants in foods	Small organic antioxidants
Vitamin A	Glutathione
B-Carotene	Cysteine
Vitamin C	Uric acid
Vitamin E	Hydroquinones
Antioxidant food additives	Antioxidant enzymes
BHT (butylated hydroxytoluene)	Glutathione peroxidases (Se)
BHA (butylated hydroxyanisole)	Glutathione transferase (no metal)
Sodium benzoate	Catalase (Fe)
Ethoxyquin	CuZn-superoxide dismutase
Propyl galate	Mn-superoxide dismutase
Fe-superoxide dismutase	(bacterial)

Table modified from Spallholz, (1989)

ESSENTIALITY OF SELENIUM AND SYNTHESIS OF THE GLUTATHIONE PEROXIDASES

Long known for its toxicity, selenium was first understood to be biologically essential for the prevention of liver cirrhosis in rats. This most important discovery, published in 1957, was made by Schwarz and Foltz (1957) in rats fed diets either lacking or low in vitamin E. Indeed, selenium may be only absolutely required in higher animals (there is no requirement in higher plants) when the lipid soluble antioxidant, vitamin E, is absent from the diet.

When dietarily present, selenium induces the synthesis of two Se-containing antioxidant enzymes, cytosolic glutathione peroxidase, (GSHPx) discovered in 1973 (Rotruck et al., 1973; Flohe et al., 1973) and phospholipid hydroperoxide glutathione peroxidase (PLGSHPx) discovered in 1985 (Urisini et al., 1982; Urisini et al., 1985). GSHPx is a 84,000 dalton cytosolic enzyme containing 4-gram-atoms of Se in tetrameric form and appears to be the major cytosolic Se-enzyme in many tissues (Tappel, 1984). PLGSHPx is a 20,000 dalton cytosolic enzyme containing 1-gram-atom of Se in monomeric form. This latter Se-enzyme is reported to be associated with biomembranes and reduces phospholipid hydroperoxides to alcohols a property not held in common with GSHPx (Ursini and Bindoli, 1987). The reactions of GSHPx, PLGSHPx and other antioxidant enzymes found in many aerobic cells are shown in Table 3. Specifically, the discovery of the reaction of PLGSHPx with membrane bound phospholipid hydroperoxide substrates provides an explanation for the sparing effect of selenium in the absence of adequate dietary vitamin E (Urisini and Bindoli, 1987; Simonarson, 1988). When vitamin E, the major membrane bound antioxidant is limiting, PLGSHPx provides the membrane with peroxidative protection. When dietary and membrane vitamin E is adequate the dietary requirement for Se and need for PLGSHPx is limited and perhaps selenium is not even required.

Table 3. Antioxidant Enzymes

Enzyme	Mineral	Reaction
Superoxide dismutase (SOD) (EC 1.15.1.1) MW 32,500	CuZn Mn Fe (bacterial)	$2O_2^- + 2H^+ \rightarrow O_2 + H_2O_2$
Glutathione peroxidase (GSHPx) (EC 1.11.1.9) MW 84,000	Se(4)	$H_2O_2 + 2GSH \rightarrow GSSG + 2H_2O$ $ROOH + 2GSH \rightarrow GSSG + ROH + H_2O$
Phospholipid Hydroperoxide Glutathione Peroxidase (PLGSHPx) (EC none assigned) MW 20,000	Se(1)	$H_2O_2 + 2GSH \rightarrow GSSG + 2H_2O$ $ROOH + 2GSH \rightarrow GSSG + ROH + H_2O$ $PLOOH + 2GSH \rightarrow GSSG + PLOH + H_2O$
Catalase (CT) (EC 1.11.1.6) MW 250,000	Fe	$2H_2O_2 \rightarrow 2H_2O + O_2$
Glutathione-S-transferases (GS-T) (EC 2.5.1.18) MW various	None	$ROOH + 2GSH \rightarrow GSSG + ROH + H_2O$

The discovery of the reaction of PLGSHPx with membrane bound phospholipid hydroperoxide substrates provides an explanation for the sparing effect of selenium in the absence of adequate vitamin E (Ursini and Bindoli, 1987; Simonarson, 1988) When vitamin E, the major membrane bound antioxidant is limiting, PLGSHPx provides the peroxidative protection. When dietary and membrane vitamin E is adequate the dietary requirement for Se and need for PLGSHPx is limited or perhaps not even required. Table modified from Spallholz (1989.)

DISTRIBUTION OF SELENIUM AND GLUTATHIONE PEROXIDASES

Selenium, the glutathione peroxidases and/or other selenium containing proteins are widely distributed in the animal kingdom (Smith and Shrift, 1979). There appears, however, to be no known requirement for selenium in higher plants but there may be a requirement in at least one green alga (Yokota et al., 1988). Tissue distribution of Se and glutathione peroxidase is probably best documented in rat tissues. Tables 4 and 5, from Behne and Wolters (1983), show the percentage of Se and GSHPx (Table 4) and the percentage of Se as GSHPx (Table 5) in rat tissues. Because these data were published in 1983 prior to the discovery of phospholipid hydroperoxidase glutathione peroxidase (Ursini et al.,1985), there will no doubt be reanalysis and perhaps redistribution of the assigned forms of selenium in these tissues.

Table 4. Distribution of Selenium and Selenium-dependent Glutathione Peroxidase Activity in Tissues of the Rat[1]

Tissue	Se Amount[2] %	GSHPx Activity[3] %
Muscle	39.8	6.1
Liver	31.7	65.6
Erythrocytes	0.9	21.2
Plasma	7.9	2.1
Kidneys	7.5	2.0
Skeleton	3.1	—
Lungs	1.2	0.8
Spleen	1.0	1.3
Pancreas	0.7	0.2
Heart	0.7	0.5
Brain	0.5	<0.1
Adrenals	0.1	<0.1
Thymus	0.1	<0.1

[1]Female animals, body mass 272 ± 14 g.
[2]Percent of total amount of Se in tissues analyzed.
[3]Percent of total GSHPx in tissues analyzed: determined in the soluble tissue fractions after centrifugation of the homogenate at 10,000 x.g.
From Behne and Wolters (1983).

Selenium is found in most all human tissues (Dickson and Tomlinson, 1967; Hojo, 1987) but even less is known of the amount and distribution of the glutathione peroxidases (Table 6). The major access to studies of selenium and glutathione peroxidase in humans has been via body fluids; whole blood, plasma, saliva, milk and urine, and cells; erythrocytes and other lymphoid cells. In humans, as in animals, selenium and glutathione peroxidase(s) have major functions in many of the cells derived from the undifferentiated stem cells of bone marrow and from cells modified by the immune system (Kiremidjian-Schumacher and Stotzky, 1987).

SELENIUM AND IMMUNITY

Selenium, in addition to its specific incorporation into the glutathione peroxidase(s) as a selenocysteine residue (Sunde and Evenson, 1987) is reported to have a variety of effects on the immune system. Reported effects of selenium on immunity include expression of specific and non-specific humoral and cell-mediated immune responses. The effects of selenium on specific components of the immune system have been recently reviewed (Kiremidjian-Schumacher and Stotzky, 1987; Peretz, 1988). Whereas much as 33-40 percent of all rat Se is present as GSHPx (Behne and Wolters, 1983), GSHPx is not uniformly distributed in tissues. Its distribution is even less well-known for individual cells other than erythrocytes (Hafeman et al., 1974). Nevertheless, GSHPx(s) and its antioxidant role in the elimination of inorganic and

148

organic hydroperoxides (Sies, 1987; Chow, 1979) must be a predominant role for selenium in cells of the immune system as well as in tissues.

The remaining focus of this Chapter is on the distribution of Se and/or GSHPx in specific cells of the immune system (Figure 2). A major function of selenium in immune cells and especially in cells that elicit the phagocytic respiratory burst (Babior et al., 1988 ; Halliwell et al., 1988) would seem to be the control of excessive production of peroxidative substrates such as H_2O_2 (Figure 3).

Table 5. Percentage of the Tissue Selenium Bound to the GSHPx[1]

Tissue	Se bound to GSHPx[2]
	%
Erythrocytes	82
Liver	63
Spleen	41
Lungs	22
Heart	20
Adrenals	18
Kidneys	11
Pancreas	11
Plasma	9
Thymus	7
Muscle	5
Brain	3
Testes[3]	<1

[1]Tissue GSHPx activity determined in the soluble tissue fractions after centrifucation of the homogenate at 10,000 x g.
[2]Female rats, body mass 272 ± 14 g.
[3]Male rats, body mass 335 ± 22 g. From Behne and Wolters (1983).

Table 6. Selenium Content of Tissues other Than Blood.

Tissue	Selenium content per gram of whole tissue (µg)	
	Infant	Adult
Stomach	0.19	0.17
Liver	0.34	0.39
Pancreas	0.05	0.13
Spleen	0.37	0.27
Kidney	0.92	0.63
Intestine	0.31	0.22
Heart	0.55	0.22
Lung	0.17	0.21
Artery	0.27	0.27
Muscle	0.31	0.40
Fat	0.09	0.12
Trachea	0.14	0.24
Gonad	0.46	0.47
Thyroid gland	0.64	1.24
Brain	0.16	0.27
Adrenal gland	0.21	0.36
Lymph node	0.26	0.10

From Dickson and Tomlinson (1967)

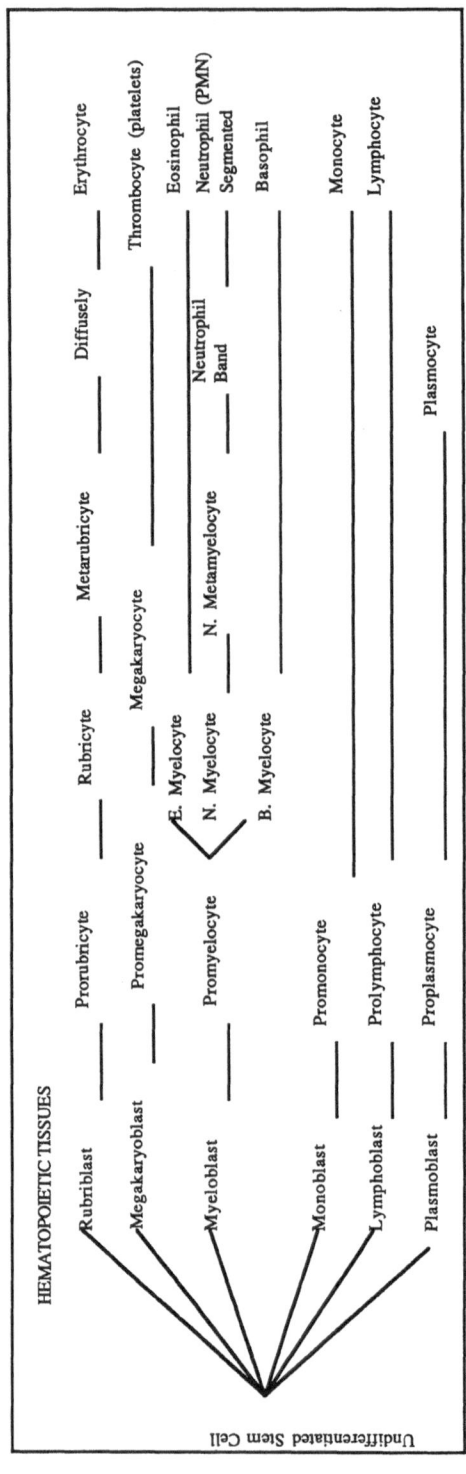

Figure 2. Differentiation of Blood and Lymphoid Cells, From Diggs et al. (1978).

Selenium and glutathione peroxidase are normally found in cells circulating in blood and in most lymphoid cells, B-, T-cells and macrophages.

150

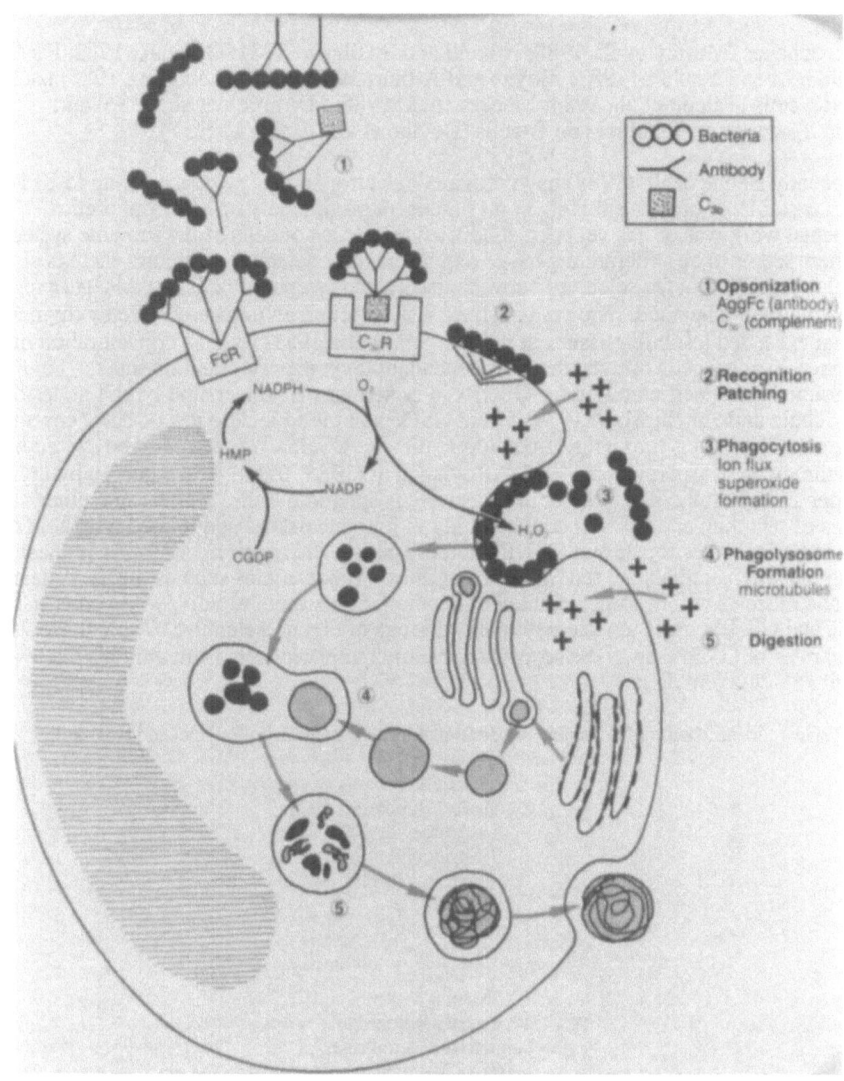

Figure 3. Phagocytosis and the Respiratory Burst (Schematic) From Sell, (1987).

SELENIUM AND GLUTATHIONE PEROXIDASE IN PHAGOCYTIC LYMPHOID CELLS

A large body of evidence points to the accumulation, presence and retention of Se and/or glutathione peroxidase in phagocytic and other lymphoid cells (Takashashi, 1986). Selenium and/or GSHPx is found in the major reticuloendothelial tissues, including bone marrow, thymus, spleen and pharangeal and mesenteric lymph nodes (Brown and Burk, 1973; Paynter, 1979). Selenium and/or glutathione peroxidase is incorporated or found in cells derived from undifferentiated bone marrow stem cells and cells modified by the immune system. In addition to the GSHPx found in erythrocytes Se and/or GSHPx is found in thrombocytes (platelets) (Bryant and Bailey, 1980; Levander et al., 1983; Levander et al., 1983; Christian, 1981; Doni et al., 1987; Kasperek et al., 1982; Menzal et al., 1983; Doni et al., 1987; Takahashi and Cohen, 1986; Thomson et al., 1987), eosinophils and neutrophils (Serfass and Ganther, 1975; Serfass and Ganther, 1976; Lindh et al., 1984; Johansson et al., 1983; Johansson and Lindh, 1984; Urban and Jarstrad, 1986) polymorphonucleated neutrophils (Bass et al., 1977; McCallister et al., 1980; Aziz et al., 1984; Aziz and Klesius, 1986) lymphocytes and granulocytes (Bracci et al., 1970; Holmes et al., 1970; Berenshtein, 1973; Berenshtein, 1974; Porter et al., 1979; Scholz and Hutchinson, 1979; Jensen et al., 1980; Clausen and Tranum, 1982; Jensen and Clausen, 1983; Baker and Cohen, 1983; Karle et al., 1983; Baker and Cohen, 1983; Baker and Cohen, 1984; Zachara et al., 1984) monocytes and macrophages (Murray et al., 1980; Novelli and Spallholz, 1981; Oh et al., 1982; Parnham et al., 1983; Li and Youhui, 1987; Boyne and Arthur, 1985; Gairola and Tai, 1985) natural killer (NK) cells (Talcott et al., 1984; Meeker et al., 1985; Dimitrov et al., 1986) and cytotoxic-T-supressor, T-helper and B-cells (Boylan et al., 1988) (Table 7).

Recently Behne et al. (1988) have provided electrophoresis evidence for up to 13 new selenium proteins or protein subunits in rat tissues. Unfortunately no electrophoretic experimental work was or has yet been conducted on tissues or cells of the immune system. While there seems to be a hierarchial tissue requirement for selenium-proteins, there is also a hierarchial requirement for selenium between tissues and lymphoid cells. In this latter regard some examples can be cited. Brown and Burk (1973) found an uptake of ^{75}Se by thymus tissue that exceeded testicular tissues in the rat. Spallholz (1981) found a concentration of ^{75}Se in peritoneal exudate cells of mice that was dependent upon dietary selenium intake. Paynter (1979) found high specific activity of GSHPx in bone marrow, spleen and lymph nodes of sheep. Scholz and Hutchinson (1979) found GSHPx in blood leukocytes of dairy cows to be 4-fold greater than in erythrocytes. In children, platelet GSHPx was found to be 5-fold higher than erythrocytes (Kasperek et al, 1982). Boylan et al (1988) found ^{75}Se taken up by thymic T-4 helper and cytotoxic-T-supressor cells as well as splenic B-cells. . These and other examples of selenium and GSHPx concentrating in immune tissues and related lymphoid cells suggest important roles for Se and GSHPx in these cells. The requirement for Se in these cells is again seen when animals are fed Se-deficient diets. Major organs such as the liver, heart and kidney and muscles will become depleted of Se and GSHPx quite rapidly, while other tissues retain Se and GSHPx more tenaciously in the absence of dietary selenium. Table 8 shows retention rates of GSHPx and ^{75}Se reported for some lymphoid cells from animals fed low or essentially Se-deficient diets.

Table 7. Identification of Selenium and/or Glutathione Peroxidase in Reticuloendotheial Cells.

B-Cell
Cytotoxic-T-suppressor
Erythrocyte
Eosinophil
Lymphocytes (PMN)
Monocyte
Neutrophil
Platelet
T-helper
Natural Killer
Peritoneal Macrophage
Pulmonary Macrophage
Aveolar Macrophage

Many effects of selenium may be simultaneously operating in cells of the immune system (Kiremidjian-Schumachear and Stotzky, 1987; Peretz, 1988). The most characterized and best understood role for selenium in immune cells is its role as an antioxidant in the GSHPx(s), a function common to most cells. Among phagocytic cells, endogenous production of O_2- and H_2O_2 as well as other free radicals during the respiratory burst (Babior et al., 1988) is generated for cytotoxic killing of the engulfed specie (Figure 3). Vitamin E, vitamin C, superoxide dismutase and GSHPx are thought to protect phagocytic cells from self-destruction by free radical peroxidation. Parnham et al. (1983) has shown that reduced GSHPx activity in macrophages results in enhanced release of H_2O_2 following zymosan induction of the respiratory burst. Enhanced H_2O_2 release is directly attributable to reduced GSHPx activity in macrophages and other phagocytic cells. Our own work, along similar experimental procedures, using zymozan stimulated resident mouse macrophages shows that indeed more H_2O_2 is produced by selenium-deficient , GSHPx deficient macrophages during the respiratory burst as measured by luminol chemiluminescence (Spallholz and Boylan, 1989). Surprisingly, O_2- production was also greater in selenium-deficient, GSHPx deficient macrophages as measured by lucigenin chemiluminescence (Figure 4). Oh et al. (1982) have also observed increased intracellular H_2O_2 production in Se-deficient macrophages. Nathan et al. (1980) found that inhibition of the cyclic redox reactions of glutathione caused macrophage-mediated cytolysis of murine tumor cells to be enhanced, results demonstrating that both normal and

Table 8. Retension of GSHPx/75-Se Activity in Lymphoid Cells.

Species	Dietary Se	Depletion Period	% of Control Animals	Cell	Reference
Rat	13 ppb	5 weeks	10-15%	PMNs	Surfass & Ganther 1975
Rat	<0.01 ppm	5 weeks	10-15%	PMNs	Surfass & Ganther 1976
Mouse	<0.01 ppm	26 weeks	~37%	Lymphocytes	Novelli & Spallholz 1981
Mouse	<0.01 ppm	26 weeks	9%	Adherent Macrophages	Novelli & Spallholz 1981
Mouse	<0.01 ppm	4 weeks	20%	PECs 75-Selenite	Novelli & Spallholz 1981
Rat	<0.01 ppm	8 weeks	8-14%	Platelets	Levander & Morris 1981
Mouse	---	9 weeks	~72%	Macrophages	Parnham et al 1983
Mouse	---	15 weeks	~85%	Macrophage	Parnham et al 1983
Mouse	---	19 weeks	~0%	Macrophage	Parnham et al 1983
Gilts	Low(?)	---	10%	PMNs	Dimitrov et al 1983
Sow	Low(?)	---	11%	PMNs	Dimitrov et al 1983
Rat	<0.01 ppm	---	6%	Macrophage	Boyne & Aruther 1985
Rat	---	12-15 wks	11%	PMNs	Boyne & Aruther 1986
Rat	<0.01 ppm	8 weeks	6%	PECs	Boyne et al 1986

From Spallholz and Stewart, (1989)

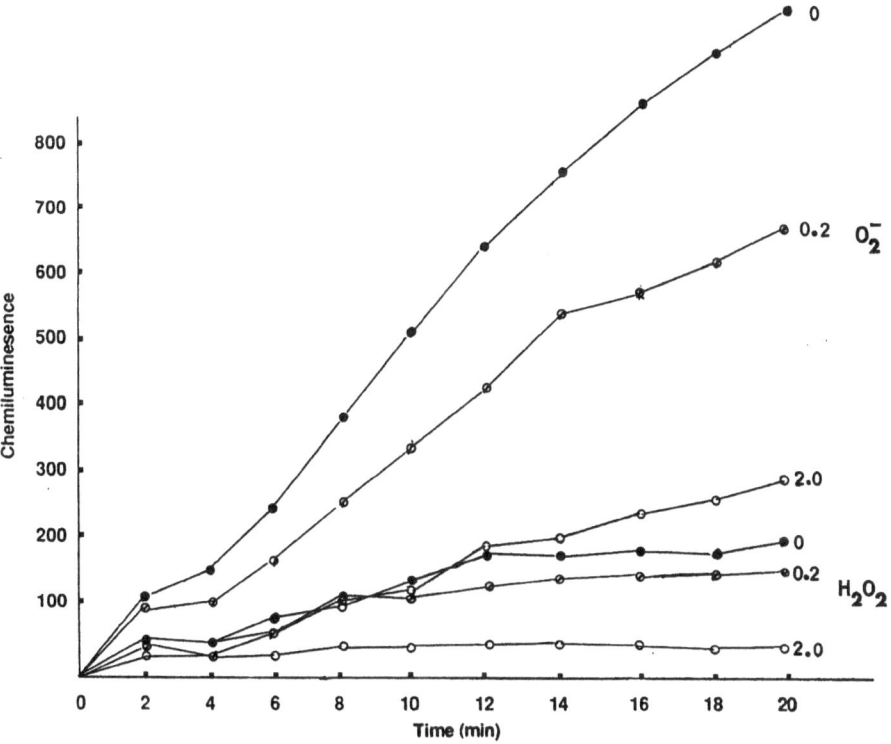

Figure 4. Chemiluminescence from resident peritoneal Balb/C macrophages fed 0, 0.2 and 2.0 ppm Se diets. Samples contained 1x 10^6 cells/tube (N=3) zymozen and lucigenin to measure O_2 or luminol to measure H_2O_2. Dietary groups are as indicated. From Spallholz and Boylan, (1989).

tumor cells use antioxidant systems for self-protection against oxidants. In addition to the antioxidant role of GSHPx in cells, selenium and/or GSHPx may have other roles in immune cells directly or indirectly related to increased endogenous release of H_2O_2. McCallister et al. (1979) have observed a loss of PMN ability to assemble microtubules and a decrease in degranulation in GSHPx-deficient cells following characteristic phagocytic stimulation. Degranulation by phagocytic cells is a necessary prerequisite to endogenous cytotoxicity.

Both normal levels of dietary Se and dietary Se in excess of that which is needed for GSHPx synthesis is known to reduce the frequency, growth and size of tumors in experimental animals and cells in tissue culture (Medina, 1986; Medina and Morrison, 1988). One immunological mechanism that may be operational is enhanced Natural Killer (NK) cell activity. NK cells do not require sensitization or antibodies to have cytotoxic selectivity towards tumor cells. Selenium given to rats in drinking water at 0.5 or 2.0 ppm exhibited enhanced NK cell cytotoxicity to tumor cells in vitro. Five ppm Se in drinking water was slightly cytotoxically suppressive to NK cells (Talcot et al., 1984). Dimitrov et al., (1986) have shown that humans ingesting 400 ug/day of sodium selenite (~160 ug Se) had increased plasma Se and enhanced NK cell activity (Figure 5) against ^{51}Cr-labeled K562 target cells. These results suggest that increases in dietary selenium may enhance cytotoxic activity of NK cells in both animals and humans. Could these NK cell effects be related to the know toxicity of selenium? Because selenium itself is toxic; is cytotoxic in its own right (Hu and Spallholz, 1982; Seko et al., 1988) and produces O_2- and H_2O_2 upon the oxidation of glutathione by selenite, (Batist et al., 1986; Seko et al., 1988) the question remains whether dietary selenium is cytotoxic in vivo in immune cells.

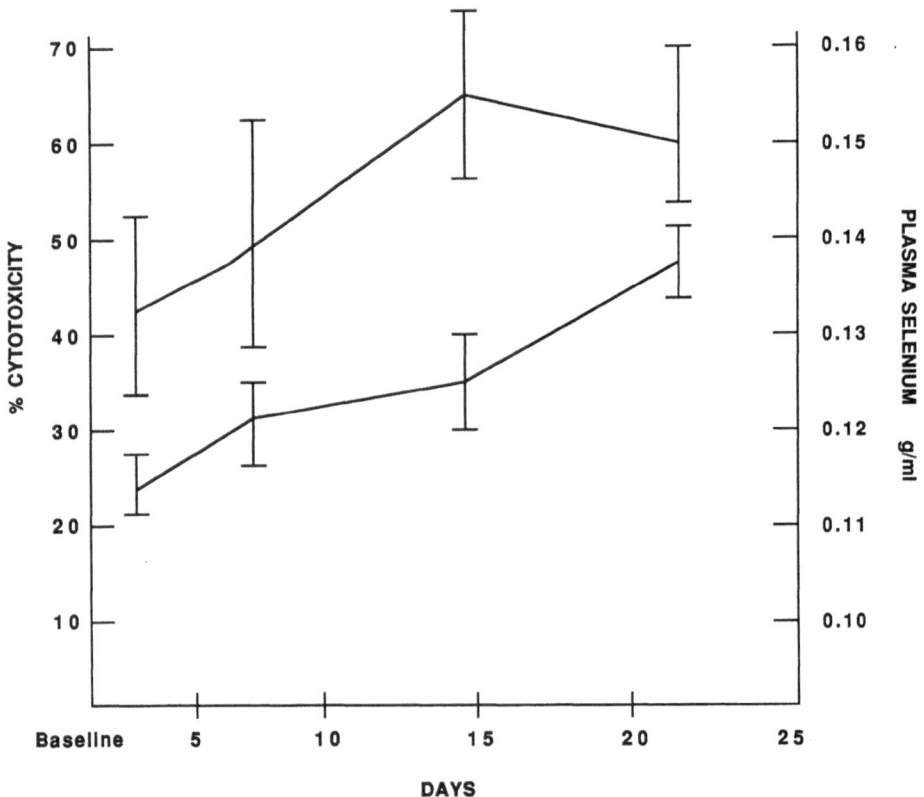

Figure 5. NK cell activity of individuals ingesting approximately 200 μg selenium from sodium selenite daily compared to plasma selenium levels. The scale on the left indicates NK cell acivity and the scale on the right plasma selenium levels. The first point represents the baseline. Each point represents mean = SD. From Dimitrov et al (1986).

Serfass and Ganther (1975) have demonstrated that Se-deficient neutrophils from rats low in GSHPx were unable to lyse *Candida albicans* following phagocytosis. Boyne and Arthur (1979, 1981) have confirmed reduced candicidal activity of Se-deficient neutrophils from cattle and demonstrated that Se-deficient mice administered *C. albicans* died more rapidly than did Se-sufficient mice (Boyne and Arthur, 1986). Arthur et al. (1981) have suggested that Se-deficient neutrophils are less able to produce the hydroxyl ($\cdot$OH) free radical.

Selenium within humans and animals plays a role in immune cells, platelets, lymphocytes, etc., in the requirement for GSHPx in the arachidonic acid cascade leading to production of prostaglandins; thromboxane, leukotrienes and other mediators of inflammatory and immune reactions. Cells devoid of Se and GSHPx are known to reduce the conversion of the arachidonic acid cyclooxygenase product, prostaglandin G_2 (PGG_2) to the corresponding alcohol, PGH_2 (Masukawa et al., 1983; Schoene, 1985; Gairola and Tai, 1985; Robinson et al., 1988).

Multiple functions of Se, GSHPx and PLGSHPx remain to be elucidated in phagocytic and non-phagocytic cells of the immune system. In time, it is likely that further understanding of the roles of selenium and other antioxidants both in protecting cells and in control of the products of the respiratory burst will provide for less disease, enhanced disease resistance and longevity through dietary manipulation of essential antioxidant nutrients.

REFERENCES

Allen, R.C., 1986, Phagocytic Leukocyte Oxygenation Activities and Chemiluminesence: A Kinetic Approach to Analysis, **Methods in Immunology,** 133:449.

Arthur, J.R., Boyne, R., Hill, H.A.O., and Ikolow-Zubkowska, M.J., 1981, The Production of Oxygen-derived Radicals by Neutrophils from Seleium-deficient Cattle, 1981, **FEBS Letters** 135:187.

Aziz, E.S., Klesius, P.H., and Frandsen, J.C., 1984, Effects of Selenium on Polymorphonuclear Leukocyte Function in Goats, **Am. J. Vet. Res.,** 45:1715.

Aziz, E.S., and Klesius, P.H., 1986, Effect of Selenium Deficiency on Caprine Polymorphonuclear Leukocyte Production of Leukotriene B4 and its Neutrophil Chemotactic Activity, **Am. J. Vet. Res.,** 47:426.

Babior, B.M., Curnutte, J.T., and Okamura, N., 1988, The Respiratory Burst Oxidase of the Human Neutrophil, *in*: **"Oxygen Radicals and Tissue Injury,"** FASEB Publisher.

Baker, S.S., and Cohen, H.J., 1983, Altered Oxidative Metabolism in Selenium-Deficient Rat Granulocytes, **J. Immunol.,** 130:2856.

Baker, S.S., and Cohen, H.J., 1984, Increased Sensitivity to H_2O_2 in Glutathione Peroxidase Deficient Rat Granulocytes, **J. Immunol.,** 114:2003.

Bass, D.A., DeChatlet, L.R., Burk, R.F., Shirley, P., and Szejda, P., 1977, Polymorphonuclear Leukocyte Bactericidal Activity and Oxidative Metabolism During Glutathione Peroxidase Deficiency, **Infect. Immun.,** 18:78.

Bates, G., Katki, A.G., Klecker, R.W., Jr., and Myers, C.E., 1986, Selnium-Induced Cytotoxicity of Human Leukemia Cells: Interaction with Reduced Glutathione, **Cancer Research,** 46:5482.

Behne, D., and Wolters, W., 1983, Distribution of Selenium and Glutathione Peroxidase in the Rat, **J. Nutr.,** 113:456.

Behne, D., Hilmer, H., Scheid, S., Gessner, H., and Elger, W., 1988, Evidence for Specific Selenium Target Tissues and New Biologically Important Selenoproteins, **Biochim. Biophy. Acta.,** 966:12.

Berenshtein, T.F., 1973, Change in the Phagocytic Activity of Leukocytes Under the Effect of Selenium and Vitamin E, **VINITI,** 7513.

Berenshtein, T.F., Effect of Selenium and Vitamin E on Phagocytic Activity of Leukocytes, **Vestsi Akad. Navuk B. SSR, Ser Biyal Navuk,** 4:123.

Boylan, L.M., Larsen, H.L. and Spallholz, J.E., 1988, Incorporation of [75]Selenium into Immune Cells in the Selenium Deficient Mouse, **Federation J.,** 2, A1621.

Boyne, R. and Arthur, J.R., 1979, Selenium-Induced Cytotoxicity of Human Leukemia Cells: Interaction with Reduced Glutathione, **Cancer Research,** 46:5482.

Boyne, R., and Arthur, J.R., 1979, Alterations of Neutrophil Functions in Selenium-Deficient Cattle, **J. Comp. Pathol.,** 89:151.

Boyne, R. and Arthur, J.R., 1981, Effects of Selenium and Copper Deficiency on Neutrophil Function in Cattle, **J. Comp. Pathol.**, 91:271.

Boyne, R., and Arthur, J.R., 1985, The Effects of Selenium Deficiency on the Function of Neutrophils in Cattle and Peritoneal Macrophages in Rats, *in*: **"Trace Elements in Man and Animals,"** C.F. Mills, I. Bremner and J.K. Chesters, eds., Commonwealth Agricultural Bureaux, Aberdeen, Scotland.

Boyne, R. and Arthur, J.R., 1986, The Response of Selenium-Deficient Mice to *Candida albicans* Infection, **J. Nutr.**, 116:816.

Bracci, R., Calobri, G., Bettini, F. and Prince, R., 1970, Glutathione Peroxidase in Human Leukocytes, **Clin. Chim. Acta**, 29:345.

Brown, D.G. and Burk, R.F., 1973, Selenium Retention in Tissues and Sperm of Rats Fed a Torula Yeast Diet, **J. Nutr.**, 102:102.

Bryant, R.W., Bailey, J.M., 1980, Altered Lipoxygenase Metabolism and Decreased Glutathione Peroxidase Activity in Platelets from Selenium-Deficient Rats, **Biochem. Biophys. Res. Commun.**, 92:268.

Chow, C.K., 1979, Nutritional Influence on Cellular Antioxidant Defense Systems, **Am. J. Clin. Nutr.**, 32:1066.

Clausen, J., and Tranum, J, 1982, Kinetics of Selenite Uptake by Mononuclear Cells from Peripheral Human Blood, **Biol. Trace Elem.**, 4:245.

Dickson, R.C., and Tomlinson, R. H., 1967, Selenium in Blood and Human Tissues, **Clin. Chim. Acta,** 16:311.

Diggs, L.W., Sturm, D. and Bell, A., 1978, *in*: **"The Morphology of Human Blood Cells,"** Abbott Laboratories, Publisher, North Chicago, IL.

Dimitrov, N.V. , Charamella, L.J., Meyer, C.J., Stowe, H.D., Ku, P,.K., and Ullrey, D.E., 1986, Modulation of Natural Killer Cell Activity by Selenium in Humans, **J. Nutr., Growth Cancer**, 3:193.

Doni, M.G., Falanga, A., Delami, F., Vicenzi, E., Tomasiak, M., and Donati, M.B., 1984, The Effect of Vitamin E or Selenium on the Oxidant-Antioxidant Balance in Rats, **Br. J. Exp. Pathol.**, 65:75.

Doni, M.G., Avventi, G.L., Bonadiman, L., and Bonacorso, G., 1981,Glutathione Peroxidase, Selenium and Prostaglandin Synthesis in Platelets, **Am. J. Physiol.**, 240:H800.

Flohe, L., Gunzler, W.A. and Schock, H.H., 1973, Glutathione Peroxidase: A Selenoenzyme, **FEBS Letters**, 32:132.

Gairola, C., and Tai, H.H., Selective Inhibition of Leukotriene B4 Biosynthesis in Rat Pulmonary Alveolar Macrophages by Dietary Selenium Deficiency, **Biochem. Biophys. Res. Commun.**, 132:397.

Hafeman, D.G., Sunde, R.A. and Hoekstra, W.G., 1974, Effects of Dietary Selenium on Erythrocyte and Liver Glutathione Peroxidase in the Rat., **J. Nutr.**, 104:580.

Halliwell, B., Hoult, J. R., and Blake, D.R., 1988, Oxidants, Inflammation and Antiinflammatory Drugs, **FASEB J.**, 2:2867.

Hawks, W.C., 1981, Selenium Metabolism in Rats: Synthesis of Glutathione Peroxidase and Other Selenocysteyl Proteins, **Diss. Abstr. Int. B.**, 41:3420.

Hojo, Y., 1987, Selenium and Glutathone Peroxidase in Human Saliva and Other Human Body Fluids, **The Science of the Total Environment**, 65:85.

Holmes, B., Park, B.H., Malawista, S.E., Quie, P.G., Nelson, D.L., and Good, R.A., 1970, Chronic Granulomotous Disease in Females: A Deficiency of Leukocyte Glutathione Peroxidase, N. Eng. J. Med., 283:217.

Hu, M.L. and Spallholz, J.E., 1983, *In Vitro* Hemolysis of Rat Erythrocytes by Selenium Compounds, **Biochem. Pharm.**, 2:5.

Jensen, G.E., Gissell-Nielsen, G. and Clausen, J., 1980, Leucocyte Glutathione Peroxidase Acitvity and Selenium Level in Multiple Sclerosis, **J. Neuro. Sci.**, 48:61.

Jensen, G.E. and Cluasen, J., 1983, Leucocyte Glutathione Peroxidase Activity and Selenium Level in Battten's Disease, **Scand. J. Clin. Lab. Invest.**, 43:187.

Joergen, C., and Tranum, J, 1982, Kinetics of Selenite Uptake by Mononuclear Cells from Peripheral Human Blood, **Biol. Trace Elem. Res.**, 4:245.

Johansson, E., Lindh, U. and Landstrom E., 1983, The Incorporation of Selenium and Alterations of Macro-and Trace Element Levels in Individual Blood Cells Following Supplementation with Sodium Selenite and Vitamin E. A Nuclear Microprobe Application, **Biol. Trace Elem. Res.**, 5:433.

Johansson, E. and Lindh, U., 1984, Uptake and Distribution of Minerals and Trace Elements in Human Blood Cells When Sodium Selenite and Vitamin E are Supplemented, **Nutr. Res. Suppl. I,** 156.

Karle, J.A., Kall, F.J., and Shrift, A., 1983, Uptake of Selenium-75 by PHA-stimulated Lymphocytes. Effect on Glutahione Peroxidase, **Biol. Trace Elem. Res.,** 5:17.

Kasperek, K., Lombeck, I., Kiem, J., Iyengar, G.V., Wang, Y.X., Feinendegen, L.E., Bremer, H.J., 1982, Platelet Selenium in Children with Normal and Low Selenium Intake, **Biol. Trace Elem. Res.,** 4:29.

Kiremidjian-Schumacher, L. and Stotzky, G., 1987, Selenium and Immune Responses, **Envirn. Res.,** 42:277.

Levander, O.A., Alfthan, G., Arvilommi, H., Greg, C.G., Huttenen, J.K., Kataja, M., Koivistoinen, P. and Pikkarainen, J., 1983, Bioavailability of Selenium to Finnish Men as Assessed by Platelet Glutahione Peroxidase Activity and Other Blood Parameters, **Am. J. Clin. Nutr.,** 37:887.

Levander, O.A., Deloach, D.R., Morris, V.C., Moser, P.B., 1983, Platelet Glutathione Peroxidase Activity as an Index of Selenium Status in Rats, **J. Nutr.,** 113:1983.

Lindh, U., Johansson, E. and Gille, L., 1984, Application of the Nuclear Microprobe to the Study of Elemental Profiles in Individual Blood Cells, **Nucl. Instrum. Methods,** 231:631.

Masukawa, T., Goto, J., and Iwata, H., 1983, Impaired Metabolism of Arachidonate in Selenium Deficient Animals, **Experientia,** 39:405.

McCallister, J., Harris, R.E., Baehner, R.L. and Boxer, L.A., 1980, Alteration of Microtubule Function in Glutathione Peroxidase-Deficient Polymorphonuclear Luekocytes, **J. Reticuloendothelial Soc.,** 27:59.

Medina, D., 1986, Mechanisms of Selenium Inhibition of Tumorigenesis, **J. Am.Col.Tox.,** 5:21.

Medina, D. and Morrison, D.G., 1988, Selenium-Mediated Inhibition of Carcinogenesis, **Phosphorous and Sulfur,** 38:3.

Meeker, H.C., Eskew, M.L., Scheuchenzuber, W., Scholz, R.W. and Zarkower, A., 1985, Antioxidant Effects on Cell-Mediated Immunity, **J. Leukocyte Biol.,** 38:451.

Menzel, H., Steiner, G., Lombeck, I., and Ohresorge, F.K., 1983, Glutathione Peroxidase and Glutathione S-Transferase Activity of Platelets, **Eur. J. Pediatr.,** 140:244.

Murray, H.W., Nathan, C.F., and Zanvil, A.C., 1980, Macrophage Oxygen-Dependent Antimicrobial Activity, **J. Exp. Med.,** 152:1610.

Nathan, C.F., Arrick, B.A., Murray, H.W., DeSantis, N.M. and Cohn, Z.A., 1980, Tumor Cell Antioxidant Defenses Inhibition of the Glutathione Redox Cycle Enhances Macrophage Mediated Cytolysis, **J. Exp. Med.,** 153:766.

Novelli, R. and Spallholz, J., 1981, Incorporation of ^{75}Se-Selenite into Peritoneal Exudate Cells of Mice Separation and Glutathione Peroxidase Assay of Adherent and Non-Adherent Cells, *in:* **"Selenium in Biology and Medicine,"** J.E. Spallholz, J.L. Martin and H.E. Ganther, eds., AVI Publishing Co., West Port, CT.

Oh, S.H., Lee, M.H., and Chung, C.J., 1982, Protection of Phagocytic Macrophages from Peroxidative Damage by Selenium and Vitamin E, **Yonsei Med. J.,** 23:101.

Parnham, M.J., Winkelmann, J., and Leyck, S., 1983, Macrophage Lymphocyte and Chronic Inflammatory Responses in Seleium Deficient Rodents. Association with Decreased Glutathione Peroxidase Activity, **Int. J. Immunopharm.,** 5:455.

Paynter, D.K., 1979, Glutathione Peroxidase and Selenium in Sheep I. Effect of Intramunal Selenium Pellets on Tissue Glutathione Peroxidase Activities, **Aust. J. Agric. Res.,** 30:695.

Peretz, A., 1989, Selenium in Inflammation and Immunity, *in:* **"Selenium in Medicine and Biology",** J. Neve and A. Favier, eds., Walter de Gruyter, New York, NY.

Porter, E.K., Karle, J.A., and Shrift, A., 1979, Uptake of Selenium-75 by Human Lymphocytes in Vitro, **J. Nutr.,** 109,:1901.

Robinson, D.R., Tateno, S., Patel, B., and Hirai, A., 1988, Lipid Mediators of Inflammatory and Immune Reactions, **J. Parenteral and Enteral Nutr.,** 12:37S.

Rotruck, J.T., Pope, A.L, Ganther, H.E., Swanson, A.B., Hafeman, D.G., and Hoekstra, W.G., 1973, Selenium: Biochemical Role as a Component of Glutathione Peroxidase, **Science,** 179:588.

Schoene, N.W., Morris, V.C., and Levander, O.A., 1986, Altered Arachidonic Acid Metabolism in Platelets and Aortas from Selenium-Deficient Rats, **Nutr. Res.,** 6:75.

Schoene, N.W., 1985, Selenium-Dependent Glutathione Peroxidase and Eicosanoid Production", *in:* **"Biochemistry of Arachidonic Acid Metabolism,"** W.E.M. Lands, ed., Martinus Nijhoff Publishing, Boston, MA.

Scholz, R.W., and Hutchinson, L.J., 1979, Distribution of Glutathione Peroxidase Activity and Selenium in the Blood of Dairy Cows, **Am. J. Vet. Res.,** 10:245.

Schwarz, K. and Foltz, C.M., 1957, Selenium as an Integral Part of Factor 3 Against Dietary Liver Degeneration, **J. Am. Chem. Soc.,** 79:3292.

Seko, Y., Saito, Y., Kitahara, J. and Imura, N., 1988, Active Oxygen Generation by the Reaction of Selenite with Reduced Glutathione *in Vitro*, *in*: **Abstract, Fourth International Symposium on Selenium in Biology and Medicine,** Tubingen, West Germany.

Seko, Y., Kitahara, J. and Imura, N., 1988, Hemolysis of Mouse Erythrocytes by Selenite *in Vitro*: A Model for Studying Cytotoxicity of Selenite, *in*: Abstract, **Fourth International Symposium on Selenium in Biology and Medicine,** Tubingen, West Germany.

Sell, S., 1987, **Immunology Immunopathology and Immunity,** Elsevier Publishing Co., New York, NY.

Serfass, R.E. and Ganther, H.E., 1975, Defective Microbicidal Activity in Glutathione Peroxidase Deficient Neutrophils of Selenium-deficient Rats, **Nature,** 225:640.

Serfass, R.E. and Ganther, H.E., 1976, Effects of Dietary Selenium and Tocopherol on Glutathione Peroxidase and Superoxide Dismutase Activities in Rat Phagocytes, **Life Sci.,** 19:1139.

Sies, H., 1967, Antioxidant Activity in Cells and Organs, **Am. Rev. Respir. Dis.,** 136:478

Simonarson, B., 1988, Glutathione Peroxidase, Selenium and Vitamin E in Defense Against Reactive Oxygen Species, *in*: **"Reactive Oxygen Species in Chemistry, Biology and Medicine,"** A. Quintahilha, ed., Plenum Publishing Company, New York, NY.

Smith, J. and Shrift, A., 1979, Phylogenetic Distribution of Glutathione Peroxidase, **Comp. Biochem. Physiol.,** 63:39.

Spallholz, J.E., 1981, Selenium: What Role in Immunity and Immune Cytotoxicity?, *in*: **"Selenium in Biology and Medicine,"** J.E. Spallholz, J.L Martin, and H.E. Ganther, AVI Publishing, Co., Westport, CT.

Spallholz, J.E., 1989, **Nutrition: Chemistry and Biology,** Prentice Hall, NJ.

Spallholz, J.E., and Stewart, J.R., 1989, Advances in the Role of Minerals in Immunobiology, **Biol. Trace Ele. Res.,** 19:129.

Spallholz, J.E., and Boylan, L.M., 1989, Effect of Dietary Selenium on Peritoneal Macrophage Chemiluminescence, **Federation J.** 3: A778.

Sunde, R.A. and Evenson, J.K., 1987, Serine Incorporation into the Selenocysteine Moiety of Glutathione Peroxidase, **J. Biol. Chem.,** 262:933.

Takashashi, K., 1986, Selenium Deficiency and Glutathione Peroxidase, **Taisha,** 23:1097.

Takashashi, K., and Cohen, H.J., 1986, Selenium-Dependent Glutathione Peroxidase Protein and Activity: Immunological Investigations on Cellular and Plasma Enzymes, **Blood,** 68:640.

Talcott, P.A., Exon, J.H., and Koller, L.D., 1984, Alteration of Natural Killer Cell-Mediated Cytotoxicity in Rats Treated with Selenium, Diethylnitrosamine and Ethylnitrosourea, **Cancer Letters,** 23:313.

Tappel, A.L., 1984, Selenium - Glutathione Peroxidase: Properties and Synthesis, **Current Top. Cell. Reg.,** 24:87.

Thomson, C.D., Steven, S.M., Van Rij, A.M., Wade, C.R., and Robinson, M.F., 1988, Selenium and Vitamin E Supplementation: Activities of Glutathione Peroxidase in Human Tissues, **Am. J. Clin. Nutr.,** 48:316.

Urban, T., and Jarstrand, C., 1986, Selenium Effects on Human Neutrophilic Granulocyte Function *in Vitro*, **Immunopharm.,** 12:167.

Ursini, F., Maiorino, M., Valente, M. and Gregolin, C., 1982, Purification from Pig Liver of a Protein Which Protects Liposomes and Biomembranes from Peroxidative Degradation and Exhibits Glutathione Peroxidase Activity on Phosphatidylcholine Hydroperoxides, **Biochim. Biophys. Acta,** 710:197.

Urisini, F., Maiorino, M. and Gregolin, C., 1985, The Selenoenzyme Phospholipid Hydroperoxide Glutathione Peroxidase, **Biochim.Biophys. Acta,** 839:62.

Ursini, F. and Bindoli, A., 1987, The Role of Selenium Peroxidases in the Protection Against Oxidative Damage of Membranes, **Chem. Phys. of Lipids,** 44:255.

Yin, L. and Zang, Y., 1987, Heterogeneity of Murine Macrophage-Mediated Cytotoxicity and the Effect of Selenium on Different Macrophage Subsets, **Zhongguo Mianyixue Zazhi,** 3:139.

Yokota, A., Shigeoka, S. Onishi, T. and Kitaoka, S., 1988, Selenium as an Inducer of Glutathione Peroxidase in Low-Carbondioxide-Grown *Chlamydomonas reinhardtii*, **Plant Physiol.,** 86:649.

Zachara, B., Gromadzinska, J., Wasowicz, W., and Sklodowaka, M., 1984, The Influence of Selenium on Glutathione Peroxidase Activity in Rabbit Reticulocytes *in Vitro*, **Bull. Pol. Biol.,** 32:111.

EPILOGUE

Robert P. Tengerdy

Department of Microbiology
Colorado State University
Fort Collins, CO 80523

The central theme of this symposium was the contribution of antioxidant nutrients to immune functions and disease protection. The effect of antioxidant vitamins A, C and E, the provitamin beta carotene, the key antioxidant micronutrients, Zn, Cu and Se, polyunsaturated lipids and natural antioxidants in human milk was surveyed on cells of the immune and phagocytic systems and on the biochemical reactions that underlie their functions. The papers presented at this symposium are a collection of key examples of ongoing research in this area by representative laboratories. The intention was to collect these results in one volume, to find the common thread in this broad and divergent area of investigations, to point out the interaction among the antioxidant nutrients and other nutrients, and to show the direction this line of research is heading to.

In these days, when the concern for proper nutrition has passed the walls of research laboratories and has reached into the public consciousness, the clear separation of established facts from conjectures, fads, or exaggerated claims and expectations has a paramount importance. The key to proper nutrition is a proper balance between the various elements of the diet; protein calorie, lipid, vitamin and mineral nutrition. This was emphasized in the keynote address by Dr. Chandra. The nutritional status of an individual may have predictive value for immunocompetence and disease protection and vice versa tests of immunocompetence may be used as functional indices of the nutritional status.

The main role of the antioxidant nutrients in a proper diet is the prevention of oxidative damage to cells and their biochemical and physiologic functions. Antioxidant nutrients counteract free radical and other oxidative damages to cell membranes, thus protecting cells. The cells involved in immune responses are among the most rapidly differentiating and proliferating cells in vertebrate animals. Such dividing and transforming cells are particularly prone to oxidative damage, hence the need for increased protection by antioxidant nutrients.

Bendich reviewed the role of antioxidant nutrients in immune responses. Antioxidant nutrients, supplemented to diets, may enhance lymphocyte proliferation, humoral antibody production and various manifestations of cell mediated immunity. Conversely, deficiency in antioxidant nutrients impair immune responses and may increase the risk of certain forms of cancer. The interaction among antioxidant nutrients modulates immune responses. The antioxidant vitamins, vitamin A, C, E and the provitamin beta carotene may interact in preventing lipid peroxidation in cell membranes. Vitamins A and E are involved in the modulation of prostaglandin, thromboxane and leukotriene biosynthesis and activity and in the modulation of the biosynthesis and activity of other mediators of immune responses. Beta carotene may have a special role in cancer prevention.

Dietary lipids, especially PUFA, modify the effects of antioxidant vitamins. Fernandes reported that in mouse diets rich in omega-3-fatty acids, such as in fish oil, lipid peroxidation was accelerated in vitamin E depleted membranes, affecting particularly the activity of natural killer cells, one of the main nonspecific defensive cells against infectious diseases and cancer. The vitamin E requirement for optimal defensive function increases with the PUFA content of the diet.

Key elements of nonspecific and specific defenses against microbial infections are the phagocytic cells, the polymorphonuclear cells (PMN) and mononuclear cells (macrophages). Boxer's presentation focused on the effect of antioxidant nutrients on these cells. For microbicidal activity, these cells rely on the respiratory burst, a strong oxidative reaction. At the same time these cells need antioxidant protection for proper cellular development and function. This seeming contradiction may be resolved by properly balancing the diet in antioxidant nutrients. When a microbial infection stimulates PMN, the resulting respiratory burst damages not only the infectious agent but the host cells as well. Antioxidant nutrients such as alpha-tocopherol or the synthetic antioxidant 2,3 dihydroxy benzoic acid may protect the host cell from this damage, while allowing the proper functioning of the PMN. The slight decrease in microbicidal activity of PMN in alpha-tocopherol repleted human volunteers was more than compensated by the increased membrane stability and increased phagocytic activity of PMN. Alpha-tocopherol also restored normal phagocytic activity in glutathione synthetase deficient patients. Excessively high doses of alpha-tocopherol, however, may lead to selective inhibition of clearing immune complexes by PMN.

The enzymes participating in the respiratory burst of phagocytes and in the protection of membranes from lipid peroxidation require metals or selenium as integral constituents, cofactors or activators. Hence a balanced mineral nutrition is critical in body defenses against infectious diseases and probably against cancer too.

Selenium is part of glutathione peroxidases and an activator of vitamin E in cell membranes. Spallholz reviewed the antioxidant role of selenium in protecting cell membranes from lipid peroxidation and thereby enhancing immune responses. The membrane bound phospholipid hydroperoxide glutathione peroxidase may play a regulatory role in this process. Selenium also enhances the phagocytic activity of PMN and the antitumor activity of NK cells.

Fraker reviewed the role of Zn in body defenses. Zn deficiency apparently reduces the number of lymphocytes available for immune defense. Zn deficiency strongly impairs the phagocytic ability of mononuclear cells, by reducing the efficiency of the respiratory burst. A complex interaction of metalloenzymes, containing Zn, Cu and Fe, or activated by these metals, regulates a chain of oxidative enzyme reactions, critical in phagocytic activity and probably in immune functions.

Copper is another key micronutrient for proper body defenses, as was pointed out by Prohaska. Copper contributes to antioxidant action as part of key cuproenzymes, regulates mediators of inflammatory responses, such as the vasoactive amines, through dopamine beta-monooxygenase, and enhances the energy metabolism of the rapidly proliferating, thus energy requiring, defensive cells through cytochrome C oxidase.

Due to complex interactions among these micronutrients, and with other nutrients, a carefully balanced diet is necessary for optimal defensive activity. The human milk is a good example for a balanced antioxidant diet, as was presented by Goldman. It contains nonspecific antiinflammatory agents, such as lysozyme, complement inhibitors, prostaglandins, lactoferrin, catalase, beta carotene, cystein, ascorbate, glutathione peroxidase and alpha tocopherol, as well as specific IgA antibodies. The lipid fraction of human milk inhibits superoxide generation by PMN.

Antioxidant nutrients may have a special role in tumor defenses. Schwartz has demonstrated the antitumor activity of beta carotene. Beta carotene administered either orally or locally to a hamster cheek pouch tumor either inhibited tumor growth or caused regression of the tumor. At the site of the regressed oral carcinoma a high concentration of tumor necrosis factor alpha and beta have been found. Beta carotene and perhaps other antioxidants may be localized to a tumor site in a delivery vehicle of a liposome. This way, possible toxic responses to the whole organism by high levels of antioxidants may be avoided. In the regressing or inhibited tumors, beta carotene and other carotenoids, such as canthaxanthine and 13-cis retinoic acid stimulated the infiltration of mononuclear cells and mast cells to the tumor site, but fewer PMNs, reversing the picture with progressing tumors. It is plausible that these events are associated with an effective tumor immunity.

The need for more effective delivery systems for some antioxidant nutrients were also emphasized by Tengerdy. In dietary supplementation only a small fraction of the antioxidant nutrient reaches the intended target organ or cells. Administering vitamin E in a water-in-oil type adjuvant, mixed with bacterial vaccines, gave far greater protection against specific bacterial infections than dietary supplementation of vitamin E. In the adjuvant, the vitamin E is in an oily depot at the site of a subcutaneous injection, in close contact both with the antigen and the cells of a local inflammatory and immune reaction. A far greater effective concentration of vitamin E may be achievable in an adjuvant than is either practical or desirable by dietary administration.

Age is an important modulator of the protective effect of antioxidant nutrients. Meydani presented data on the effect of antioxidant nutrients on defense functions in aged populations of mice and human volunteers. The main indicators of immunologic effectiveness decrease with age, such as antibody production, mitogen or antigen stimulation of lymphocytes, other manifestations of cell mediated immunity or nonspecific defenses. The serum level of lipoperoxides rises with age. At the same time, antioxidant levels, vitamin C, Fe, glutathione and superoxide dismutase decrease with age. The accumulated peroxidative damage to cells and organs probably contributes to the degeneration of the immune system. The supplementation with the antioxidant nutrient, vitamin E, resulted in improved immune functions in the elderly.

The papers presented at this symposium demonstrated the benefit of antioxidant nutrients in maintaining or enhancing the efficiency of nonspecific and specific defensive mechanisms of the human and animal body. The key steps in antioxidant protection of the immune system as well as improved phagocytosis by PMNs and mononuclear cells have been elucidated. The participation of antioxidant nutrients in key enzyme reactions of these events has been demonstrated. The need for balancing oxidative and antioxidative defensive reactions is recognized and the case was well documented. The key target cells of the antioxidant nutrients for improved defensive functions have been identified. These are the lymphocytes, polymorphonuclear and mononuclear phagocytic cells, and natural killer cells. The antitumor effect of antioxidant nutrients have been linked to possible immune and other defensive mechanisms.

The mechanisms by which these antioxidant nutrients enhance immunity and other defensive mechanisms are intensively studied, but much more work is still ahead. The interaction among antioxidant nutrients and other nutrients is not yet well understood, the available data are sporadic and sometimes controversial. The importance of the proper dosage of these nutrients is emphasized and the need for better delivery systems is recognized. Liposomes and adjuvants are a good start in this direction. The symposium summarized the current status of research in this area and pointed toward new directions in future research and development.

PARTICIPANTS

Adrianne Bendich
Clinical Research Coordinator
Hoffmann-LaRoche Inc.
340 Kingsland St.
Nutley, NJ

Laurence A. Boxer
Division of Pediatric Hematology/Oncology
CS Mott Childrens Hospital
University of Michigan
P. O. Box 0238
Ann Arbor, MI 48109

Ranjit Kumar Chandra
Departments of Pediatrics and Medicine
Health Sciences Centre
Memorial University of Newfoundland
St. John's, Newfoundland A1B 3V6, Canada

Gabriel Fernandes
Department of Medicine and Physiology
The University of Texas Health Science Center
San Antonio, TX

Pamela J. Fraker
Department of Biochemistry
Michigan State University
East Lansing, MI 48824

Armond S. Goldman
The University of Texas Medical Branch
Galveston, TX

Simin Nikbin Meydani
USDA Human Nutrition Research Center on Aging
Tufts University
Boston, MA

Marshall Phillips
USDA-Agricultural Research Service
National Animal Disease Center
Ames, IA 50010

Joseph R. Prohaska
Departments of Biochemistry and Medical Microbiology
  and Immunology
University of Minnesota
Duluth, MN 55812

Joel L. Schwartz
Harvard School of Dental Medicine
Department of Oral Pathology and Oral Medicine
188 Longwood Ave.
Boston, MA 02115

Julian E. Spallholz
Texas Tech University
Institute for Nutritional Sciences and
  Center for Food and Nutrition
Lubbock, TX 79409

Robert B. Tengerdy
Department of Microbiology
Colorado State University
Fort Collins, CO 80523

INDEX

Carcinogen, 78, 85
Carcinogenesis, 78, 79
Carcinoma, 41, 49, 84, 88, 90, 92, 96
Cardiovascular disease, 57, 66, 102
Carotenoids, 4, 7, 9, 46, 48, 79, 91, 92, 161- (see
    alpha, carotene, beta carotene, canthaxanthin,
    astaxanthin, bixin, citral)
Catalase, 3, 7, 8, 21, 23, 30, 31, 71, 97, 116, 132,
    135
Catecholamines, 2
Cattle, 8, 108, 111, 139
Cell cycling (DNA synthesis), 139
Cell mediated immune responses, 14, 15, 17, 63,
    106, 143, 148, 154
    lympholysis assay, 86
    responses to tumors, 48, 119
Cell receptor functions, 108
Cellular respiration, 1
Cereals, 126
Cerebellum, 60
Cerious chloride, 25
Cerium perhydroxide, 25
Ceruloplasmin, 5, 125, 129, 140, 141, 142, 145,
    146 - (see copper)
Chaga's disease, 113, 114
Chemiluminescence, 36, 154
Chemotactic peptides, 24, 25
Chemotaxis, 6, 23, 28, 36, 44
Chickens, 8, 42, 59, 111
Chloramine, 20
Cholesterol, 140
Chronic diseases, 57
Chronic granulomatous disease, 24, 26, 28, 30, 33
Chronic inflammatory response, 7, 39
Citral, 47
Coagulation system, 70, 71
Cobalt, 119
Cobra venom (naja naja) factor, 29
Coconut oil, 39
Collagen, 43
Colon, 48
Colony stimulating factors, 15
Colostrum - (see human milk)
Complement, 7, 14, 16, 23, 31, 32, 37, 45, 72, 76,
    135
    C1q, 46
    C3 activites, 25, 26
    C3 levels, 16, 67
    C5 related chemotactic activity, 32
    C5a, 24, 75
    Fc and complement receptors, 33, 88, 119
Concanavalin A (Con A), 45, 58, 62, 63, 86, 88,
    135, 142 - (see lymphocyte proliferation
    responses to mitogens)
Conjugated dienes, 30
Copper, 7, 8, 13, 15, 116, 119, 125, 126, 129,
    138, 141, 160- (see ceruloplasmin)
    chronic copper deficiency, 144, 145
    cuproenzymes, 130, 142
    cuproproteins, 130, 131, 146
    deficiency, 139, 141, 144
    deficient mice, 136, 138, 140, 142-145
    deficient infants, 132
    dietary deficiency, 130, 131, 134, 136, 137,
    143, 144
    environmental deficiency, humans, 130
    functional deficiency, 137, 145

Copper (continued)
    genetic deficiency, 130, 132, 134, 136, 144
    metabolism, 146
    status, 139
    supplementation, 131
    thymus level, 142
Copper-zinc superoxide dismutase (Cu, Zn-SOD,
    EC 1.15.1.1), 96, 123, 130, 135, 137, 142,
    145, 146
Corn oil, 41, 63
Corticosteroids, 30, 113, 146
Corynebacterium bovis, 106
Cow's milk, 72, 77
Coxsackie virus, 103
Crohn's inflammatory bowel disease, 13
Cyclic AMP, 22, 146
Cyclic GMP, 63
Cyclooxygenase, 31, 40, 64
Cyclooxygenase inhibitor, 59
Cysteine, 71, 72
Cytochrome c oxidase (EC 1.9.3.1), 77, 130, 135,
    137, 142, 145, 146, 153
Cytochromes, 22, 23, 25
Cytokines, 77
Cytoplasmic SOD, 119
Cytotoxic agents, 64
Cytotoxic T lymphocyte activity - (see
    T lymphocytes)

Dehydrogenases, 3
Delayed cutaneous hypersensitivity (DCH), 14-17,
    44-48, 58-60, 86, 119, 143
Deoxynucleotidyl transferase activity, 14
Dermis, 83, 85, 86
Desferrioxamine, 21
Desmosomes, 78
Diacylgcero, 23
Diacylglycerol lipase, 116
Diarrhoeal disease, 14
Dietary corn oil (CO), 98
Dietary fat, 108
Dietary fish oil (FO), 98
Dihomogamma-linolenic acid (DHGL), 117
2, 3-Dihydroxybenzoic acid, 21, 29, 30, 160
7, 12-Dimethylbenz(a)anthracene (DMBA), 48, 78
Dimethylformamid, 119
1, 2-Dimethylhydrazine (DMH), 48
2, 4-Dinitro-7-fluorobenzene, (DNFB) 59
Dioxygen (O$_2$) - (see molecular oxygen)
Diphtheria toxoid, 62
Directed cell movement, 23
Dl-a-tocopheryl acetate, 63- (see Vitamin E)
DMBA, 79, 81, 82, 83, 88, 91, 97
DNA, 2, 22
DNA polymerase, 114
DNA synthesis, 139
Docosahexaenoic (DHA) acids, 102, 106
Dogs, 41
Domestic animals, 125, 127
Dopamine-b-monooxygenase (DBM), 141
Double blind, placebo-controlled clinical trial, 44, 63
Dyhydroxybenzoic acid, 23
Dysplasia, 85, 88

Eicosanoid production, 141
Eicosapentaenoic acid, 6, 95, 106, 152
Electron donor, 19

Hydroxylation reactions, 43
Hypercholesterolemia, 140
Hyperkeratosis, 78, 85
Hypochlorite ion, 2, 21, 22
Hypogammaglobulinemia, 126
Hypothalamic-pituitary-gonadal-thymic axis, 141
Hypothalamus, 141

Ia antigen, 140
IgA, 108, 75 - (see antibodies, secretory IgA)
IgE, 69, 70
IgG, 63, 69, 71, 105
IgG opsonized paraffin oil droplets, 23
IgM, 63, 69, 70, 105, 141
Immune complexes, 21, 22, 31
Immunization, 17
Immunodeficiency, 13
Immunoglobulins, 43, 126 - (see antibodies)
Inanition, 114
Increased resistance, 38
Indomethacin (a cyclooxygenase inhibitor), 40, 62
Infants - (see Human population groups)
Infections, 13, 37, 43, 44, 61, 65, 104
Inflammation, 2, 20, 21, 23, 33, 43, 69, 77, 83, 85,
    90, 92, 96, 142-144
Inherited phagocytosis disorders, 6
Interferons, 7, 15, 43, 48
Interleukins
    interleukin- 1 (IL-1), 7, 16, 58, 62, 66, 67, 97,
        144
    interleukin- 2 (IL-2), 15, 16, 37, 58, 62, 88,
        111, 119, 144
    interleukin- 3 (IL-3), 16
    interleukin- 4 (IL-4), 16
Intracellular killing, 23
Intracellular parasitic pathogens, 7
Intrauterine malnutrition, 14
Iron, 2, 13, 15, 20, 21, 30, 71, 106, 118, 119, 126,
    129, 130, 131, 134, 137, 138
    chelator, 26
    deficiency 13

Kallikrien, 70
Keratin, 80
Keratinocytes, 92
Keyhole limpet hemocyanin (KLH), 138
Kidney, 7, 64, 85, 131, 132
Kwashiorkor, 14

L-ascorbic acid - (see Vitamin C)
Laboratory animals, 105
Lactate/pyruvate, 140
Lactation - (see Human milk)
Lactoferrin, 15, 21, 69
Lard, 39
Leukocyte migration inhibition factor, 15
Leukocytes, 6, 8, 28, 69, 114
Leukoplakia, 80
Leukotriene, 6, 7, 40, 58 67, 72, 73, 113, 116, 157
Linoleic acid, 6, 24, 99
Lipid, 23, 24, 63, 108
    hydroperoxides, 6, 24
    lipid-indu, ced immunosuppression, 39
    lipidemia 102
    peroxidation, 5, 6, 7, 21, 40, 64, 73, 74, 95,
        97, 103, 140
    peroxides, 3, 5, 21, 30, 40, 42, 62, 64, 154

Lipooxygenase, 64
Lipophilic amines, 20
Lipopolysaccharide (LPS), 63, 125, 130, 147 - (see
    mitogen-stimulated lymphocyte proliferation)
Lipoproteins, 6, 98
Liposomes, 5, 77, 83, 88, 90, 161
Lipoxygenase, 3, 40, 44
Liver, 3, 63, 85, 102, 130, 132, 140, 149
    cirrhosis, 149
    Cu, 135
    disease, 113
    GSH Px activity, 132
    GSH transferase, 132
    mitochondrial and microsomes, 97, 106, 124
    TBA-positive materials, 102
    tocopherol, 60
Lung, 2, 3, 7, 23, 29, 30, 37, 46, 80, 85
Luteinizing hormone releasing hormone (LHRH),
    141
Lymphocytes, 6, 8, 16, 35, 40, 41, 58, 90, 92, 111,
    118, 119, 126, 144-146, 153
    activation, 73
    development, 127
    differentiation, 59
    lymphokine production, 49
    membrane, 36, 140
    proliferation response to mitogens, 8, 14, 15,
        17, 37-39, 42-49, 51, 57-64, 67, 84, 111,
        119, 130, 140-143, 145
Lymphoid organ, s 39
    atrophy, 16
    cells, 16
    lymph node, 14, 50, 114, 119, 153
    lysosomes, 2, 30, 90, 92
Lysozyme, 69, 73, 74
Lyt 1 (helper) subset, 139

Macrophages, 2, 6, 7, 35, 36, 42-49, 58, 64, 71,
    86, 88, 92, 102, 107, 1108, 114, 116, 120,
    130, 137, 138, 139, 140, 142, 160
    chemotaxis, 7, 36
    cytotoxicity, 80
    membrane receptors for Ia antigen, s 37
    migration inhibition factor, 15
    peritonial, 119, 144, 145, 147
Malabsorption, 16
Malnutrition, 13, 57
    maternal, 17
    protein-calorie, 14, 15, 17
Malonyldialdehyde, 23
Mammary gland, 72
    growth factors, 72
    mastitis, 113
Manganese, 116, 119
    superoxide dismutase, 96, 97,
Mannitol, 21, 23
Marasmus, 14
Mast cells, 69, 70, 86, 88, 91, 93
Membrane
    fluidity, 41, 102, 145
    lipid peroxidation, 38, 102
    phospholipids, 42, 62
    receptors, 96
Menkes' disease, 125, 127, 138
2-Mercaptoethanol, 36, 37, 88
2-Mercaptoethanolamine, 59, 63
Metalloenzymes, 3, 114, 162

Phospholipid liposomes, 30
Phospholipids, 24, 42 92, 116, 117, 145
Phosphorylation, 23
Phycotene (algal extract), 48
Phytohemagglutinin (PHA), 43, 62, 63, 140, 141 - (see lymphocyte proliferation response to mitogens)
Pigs, 8, 41, 60, 111, 132
Pituitary, 141
Plaque forming cells (PFC), 130
Plasma membrane, 2, 22, 24, 25
Plasminogen activator, 72
Platelets, 6, 8, 36, 72, 102
    phospholipase A2 activity, 42
Pneumocystis carinii, 14
Pokeweed (PWM), 130
Polymorphonucleated neutrophils(PMN), 6, 90, 92, 97, 106, 124, 160
Polyunsaturated fatty acids (PUFA), 5, 42, 77, 106, 110, 113, 140, 160
Poultry, 104, 111
Prealbumin levels, 45
Pregnancy, 74
Pregnancy associated alpha 2-glycoprotein, 73
Promotion stage of tumor development, 93
Prostacyclin, 141
Prostaglandin, 2, 6, 8, 42, 57, 106, 141
Prostaglandin E2 (PGE2), 42, 58, 61, 62, 63, 76, 106, 141, 146, 147, 157
Prostate, 80
Protein kinase, C 22, 23, 116
Protein metabolism, 16
Protein phosphorylation, 22
Proteolytic enzymes, 73
Pyridine nucleotides, 21, 23, 145
Pyridoxine, 15
Pyrroloquinoline quinone, 125

Radiation-induced immunosuppression, 50
Rats, 23, 35, 39, 42, 62, 122, 127, 138, 140
Reactive oxygen intermediates, 1, 6, 36-(see free radicals)
Recurrent otitis media, 28
Red blood cells, 6, 8, 35- (see erythrocytes)
Renal disease, 113
Reproduction, 60, 146
Respiratory burst, 3, 19, 20-23, 31, 116, 151, 154, 161 - (see oxygen burst)
Respiratory system, 69, 72
Respiratory tract infections, 41, 131
Reticulocytes, 35
Reticuloendothelial tissues, 151
Retinoic acid, 48, 79
    13-cis, 47, 48, 80, 81, 86, 161
    all-trans, 46
Retinol (vitamin A), 47
Retinopathy, 31
Rheumatoid arthritis, 7, 23, 98
Riboflavin, 4
Ribonucleotidyl reductase, 15
RNA, 114
Rodents, 57, 125- (see rats and mice)
Rotavirus, 71

Safety, 43, 49
Salmonella typhimurium, 15, 72, 127
Saturated fat, s 24, 39, 108

Se, 106, 109, 153-(see selenium)
    deficient macrophages, 154
    deficient neutrophil, s 157
Secretory IgA, 14, 71, 73, 74- (see IgA)
Selenium, 3, 5, 7, 8, 13, 15, 40, 63, 105, 145, 149, 150, 154, 156, 162
    macrophage levels, 154
    plasma level, 156
    thymus, 154
    thymic cells, 154
Serum vitamin levels, 39
Sheep, 8, 105, 111, 114, 139
Sheep red blood cell (SRBC) plaque forming cells (PFC,) 29, 43, 59, 143
Shigella, 71
Sickle cell anemia, 113
Silica, 36
Singlet oxygen, 1, 2, 3, 4, 6, 36, 43, 46
Skin, 43, 79
Sodium chromate-radiolabelled 51HCPC-1 tumor cells, 84
Spleen, 14, 15, 40, 58, 84, 85, 88, 129, 131, 132, 133, 135, 137, 139, 153
    cells, 8, 62, 67, 88, 90, 135, 140, 143, 152
    mononuclear phagocyte (MNP), 114
    natural killer cell ability to lyse tumor, 39
    prostaglandin E2, 41
Spontaneously hypertensive rats (SHR), 60
Squamous cell carcinoma, 85, 90
Staphylococcus aureus, 28
Stem cells, 58
Stress, 108, 111
Sulfhydryl groups, 20, 63
Sulfhydryl oxidase, 141
Sulfhydryl-oxidation reaction, 31
Superoxide dismutase (SOD), 3, 7, 8, 23, 25, 30, 32, 64, 96, 119, 121, 124 - (see copper/zinc superoxide dismutase, manganase superoxide dismustase)
    dismutases, 3
    dismutation, 3
    radical, 2, 19, 24, 26, 30, 121, 124, 128
Supplementation, 16, 47, 111
Suppressor molecules synthesized by tumors, 46, 62
Surgical patients, 17
Synovial fluid, 7
Systemic infections, 43

T and B lymphocyte proliferative responses, 41, 46-48, 140, 141, 144 - (see lymphocyte proliferation response to mitogens)
T and B lymphocytes, 8, 40, 50
T cell lipids, 8
12-Tetradecanoyl-phorbol-13-acetate (TPA), 77
T helper lymphocytes, 88- see T lymphocytes
T lymphocytes, 7, 8, 15, 17, 40, 47, 62, 67, 90, 144
    CD4+ cells, 16
    cytotoxic activity, 9, 46, 50, 84, 88, 91-93, 154
    helper T cells, 15, 50, 63,88,90,92,113,114,144
    helper/suppressor T lymphocyte ratio, 15, 43, 65
    killer cells, 16
    number, 17
    replacing factor (IL-5), 144
    suppresor, 15, 154
    thy-I positive, 88
Taurine, 20